DIGITAL INTEGRATED CIRCUITS

D1295509

Joseph E. Kasper and **Steven A. Feller** teach
at Coe College in Cedar Rapids, Iowa. Both have
had long experience in teaching digital electronics
to people of various ages and backgrounds.

DIGITAL INTEGRATED CIRCUITS

Joseph E. Kasper

Steven A. Feller

A SPECTRUM BOOK   Prentice-Hall, Inc. Englewood Cliffs, New Jersey 07632

Library of Congress Cataloging in Publication Data

Kasper, Joseph Emil, date.
  Digital integrated circuits.

  "A Spectrum Book."
  Includes index.
  1. Digital electronics.   2. Integrated
circuits.   I. Feller, Steven A.   II. Title.
TK7868.D5K38   1983   621.381'73   82-10150
ISBN 0-13-213587-6
ISBN 0-13-213579-5 (pbk.)

This Spectrum Book can be made available to businesses
and organizations at a special discount when ordered in
large quantities. For more information contact:
Prentice-Hall, Inc., General Publishing Division,
Special Sales, Englewood Cliffs, New Jersey 07632.

© 1983 by Prentice-Hall, Inc., Englewood Cliffs, New Jersey 07632.
All rights reserved. No part of this book may be reproduced in any form
or by any means without permission in writing from the publisher.
A Spectrum Book. Printed in the United States of America.

10   9   8   7   6   5   4   3

ISBN 0-13-213587-6

ISBN 0-13-213579-5 {PBK}

Page layout by Diane Koromhas
Cover design by Jeannette Jacobs
Manufacturing buyer: Barbara Frick

Prentice-Hall International, Inc., *London*
Prentice-Hall of Australia Pty. Limited, *Sydney*
Prentice-Hall Canada Inc., *Toronto*
Prentice-Hall of India Private Limited, *New Delhi*
Prentice-Hall of Japan, Inc., *Tokyo*
Prentice-Hall of Southeast Asia Pte. Ltd., *Singapore*
Whitehall Books Limited, *Wellington, New Zealand*
Editoria Prentice-Hall do Brasil Ltda., *Rio de Janeiro*

This book is dedicated to Barbara Feller and Clara Kasper for their patience and encouragement, to Clara for hours of typing, and to numerous students whose work with the text and laboratory instructions helped us make this book less imperfect than it would otherwise have been.

Contents

11

Multiplexing and Associated Ideas   117

12

Memories   126

Acknowledgments

Figure 9.9 is reprinted from "Electronic Numbers," by Alan Sobel. Copyright © by *Scientific American*, Inc., page 66. Used with permission of W. H. Freeman and Company, publishers of "Scientific American." All rights reserved.

Figure 10.16 is reprinted from "TTL Catalog Supplement from Texas Instruments," Texas Instruments Corporation, page S8–6, 1970. Courtesy of Texas Instruments, Incorporated.

Figure 11.7 is reprinted from *The Bugbook II*, by D. G. Larsen, P. R. Rony, and R. A. Braden, pages 7–8E. E & L Instruments, Inc., Derby, CT. Copyright © 1974 by Peter R. Rony. Used with permission.

Figures 11.8 and 13.8 are reprinted from *TTL Cookbook*, by Don Lancaster, pages 278–79 and 295–96. Copyright © 1974 by Howard W. Sams & Co., Inc., Indianapolis, IN. Used with permission.

Figures 13.6 and 13.7 reprinted from "More on Shift Registers" by Forest M. Mims, *Popular Electronics*, November 1980, page 108 and 109. Reprinted from *Popular Electronics Magazine*. Copyright © 1980 Ziff-Davis Publishing Company. Used with permission.

Table 14.1 is reprinted from *TTL Data Book*, National Semiconductor Corporation, page 2–109. National Semiconductor Corporation. Copyright © 1981. Used with permission.

Figures 14.5 and 14.6 and accompanying text discussions are based upon material from *Introduction to Digital Techniques* by D. I. Porat and A. Barna, page 340. John Wiley & Sons, Inc. Copyright © 1979. Used with permission.

1

An Introduction
to This Book

The General Nature of the Subject

A great electronic technological revolution is taking place in our world. The development of integrated circuit devices has opened up vast new possibilities for manufacturers, scientists, engineers, students, hobbyists, and others. The integrated circuit (IC) is becoming commonplace in our lives, and it holds great interest for a wide variety of people.

Anyone who has some idea of the complexity of discrete component transistor electronics might suppose that IC electronics is even more complex and harder to learn on the ground that IC devices can be quite sophisticated. Actually, a characteristic that distinguishes digital IC electronics from its parent subject—transistor electronics—is the ease with which it can be learned and put to use. It is not an exaggeration to state that anyone with but the barest knowledge of basic circuit ideas and no prior knowledge of transistor electronics will have no trouble in understanding digital IC electronics.

The fundamental reason for this is that ICs come in standard sealed building-block packages of transistor circuitry that you can put together easily to form more elaborate systems. The design and building of systems does not in most cases require detailed knowledge of the inner depths of the devices themselves. Only a small fraction of all the users of IC devices have such detailed knowledge.

An IC is a small rectangular block out of which pins project on which the user makes external connections. What you need to know basically is which pins are the input pins, which are the output pins, and what information the devices will produce at their output pins if you supply certain information at the input pins. It is simple to learn this but that does not indicate that the subject is therefore trivial in its consequences. On the contrary, what can be done on the basis of such information is impressive. This is an example of the whole being greater than the sum of its parts.

For Whom This Book Is Written

In writing this book we have in mind high school students, college students, computer-oriented people who want to learn something about the hardware which is used in computers, hobbyists, and, in fact, almost anyone who may be interested in learning the subject. For all these potential readers we have sought to provide a quick and efficient route to understanding and using ICs. The material in this book can be covered in a class of high school or college students in less than one semester or quarter even with a substantial part of the time given over to laboratory work. Users who are studying on their own will, we believe, be able to become functional in digital IC electronics through use of this book. Such people will benefit greatly by equipping themselves with a small amount of inexpensive apparatus and trying some experiments with actual ICs. Other readers should find the concepts discussed in this book to be useful *as concepts* even without supplementing their reading with experimental work.

As we write this book, there appears to be no other like it. There are available enormously useful books such as *TTL Cookbook* and the *CMOS Cookbook.*[1] However, these admirable books (which serious readers should obtain) are not completely basic introductions to the subject which raw beginners can use to learn the subject. Our goal has been to provide an easily read introduction to the subject so that users can then go on to use the other literature.

A few words are important about the background which we assume the user of this book to have. Readers will need to know (or perhaps to refresh their knowledge of) the basic DC circuit ideas. By this we mean the concepts of voltage, current, resistance, capacitance, power, and Ohm's law. No special knowledge of semiconductor electronics proper is assumed. With respect to mathematics, readers need only to be facile enough with algebra to be able to work with such simple relations as that contained in the statement of Ohm's law, $V = IR$, and either to be reasonably familiar with the binary number system or be willing to acquire such familiarity. For the reader who has not had an introduction to the binary number system, we include the material required in Appendices A and B.

---

[1] *TTL Cookbook,* © 1974, and *CMOS Cookbook,* © 1977, both by Don Lancaster, Howard W. Sams and Co., Indianapolis, IN.

We think it encouraging that we have led scores of people through some or all of the contents of this book. They have ranged from youngsters of junior high school age through senior citizens. The backgrounds of these people have ranged from minimal acquaintance with simple DC circuit ideas to professional experience in several branches of engineering and science. It has been gratifying to observe the rapidity with which interested persons can rise to a respectable level of competence and understanding—and to share with them their obvious sense of fun as they go along. We know from experience that learning the subject can be done, and done agreeably.

Comments About Laboratory Experimentation

Actual building of some digital IC circuits, studying their behavior and trying out variations on them is a very valuable experience. Your internalization of the meaning of what are otherwise abstract ideas will be much enhanced by constructing some circuits. Hands-on experimentation is a *great* aid in learning the subject.

Appendix C of this book offers instructions for equipping yourself at small cost with the hardware needed to build circuits and observe their functioning. At the ends of several of the chapters in this book we suggest instructive experiments which are related to the text material. The reader who may want more elaborate experiment instructions might consider using one of the published laboratory manuals which we refer to in Appendix C. However, we have made an effort to make our discussions of laboratory experiments sufficiently detailed that this book will be found to be self-contained.

Some Specific Objectives

It is quite possible to specify definite and realistic objectives which can be attained by self-study of this book or by attending a course in which this book is used as an introductory text.

• You will learn the major types of digital ICs which are readily available today and what their uses are. Logic gates, flip-flops, counters, decoder/drivers, and memory devices are a few of those with which you should be familiar and the familiarity can be acquired easily.

• You will find the rest of the world of digital IC electronics opened to you. The many books and magazine articles which are published (for the most part) assume some fundamental body of knowledge on the part of the reader. That fundamental knowledge is contained in this book.

• You can begin to delve into a more specialized area which may interest you; microcomputer hardware for example.

The following objectives depend more or less on having gained some experience in handling ICs through experimentation.

• You will develop the ability to build devices from the circuit diagrams and explanations which appear in great quantity in other books and articles. Digital voltmeters, frequency meters, clocks, timers, and a host of other possibilities exist.

• You can learn to design systems of your own. Naturally, systems which involve subtleties due to their complexity or due to special conditions such as very-high-speed requirements will present challenges to the novice, but the design of more modest digital IC devices is quite possible.

2

An Introduction
to ICs and the NAND Gate
in Particular

An Example of a Digital System:
A Pulse Counter

We start by considering the problem of counting the number of objects passing a point, such as automobiles on a street or bottles on a conveyor belt. The purpose is to show how the problem can be solved by interconnecting certain fundamental building-block units, each of which performs specific and limited functions to make the counter system.

Figure 2.1 shows the essentials of a pulse-counting system. The *pulse source* could be, for example, a photoelectric cell which generates a momentary voltage whenever it senses that an object passes by it. In the drawing it is assumed that the voltage pulses are in the form of sharp spikes although they could be of some other form. The *units counter* is fed these pulses and is to count them sequentially as follows: 0, 1, 2, 3, 4, 5, 6, 7, 8, 9, 0, 1, 2, 3, . . . , recycling after each 9 occurs. The counter is assumed to require input pulses with steep rises, sufficiently long, flat tops and steep falls if it is to function properly. A *pulse shaper* has been interposed between the pulse source and the units counter, its purpose being to convert the spikes into neat rectangular pulses of suitable shape.

The units counter feeds its output through a *decoder* to a seven-segment display unit of the kind which is familiar to almost everyone in this age of pocket calculators and digital watches. In actual IC practice, the counter output is not in an electrical form which can cause the seven-segment display unit to show the corresponding decimal digit. The decoder provides the conversion of the counter output into a suitable form.

5

Whenever the units counter completes a *decade* of counting and goes from a count of 9 back to a count of zero, it generates a pulse which the *tens counter* counts. The tens counter drives its own associated decoder and this properly illuminates the tens display unit. Thus the units and tens subsystems can show counts ranging from 00 to 99 on the two rightmost display units. In similar fashion, the tens counter drives the hundreds counter which in turn drives the thousands counter. When the total count becomes eventually 9999, the pulse from the pulse source returns the count displayed to 0000.

In the drawing the count is shown as 4892 for the sake of illustration. If a system capable of counting beyond 9999 is required, more counters, decoders, and display units can be added. (The consequent proliferation of integrated circuits and interconnecting wires can be very much reduced by more sophisticated techniques, as will be explained later.)

The system described thus far needs subsystems which perform the functions of pulse shaper, counter, and decoder. Devices which carry out these functions are of such widespread usefulness that they are manufactured in standardized forms. They are cheap, readily available, reliable, and easy to use. Such basic devices are examples of the integrated circuit building-blocks out of which more complex systems can be constructed. A counter system even as simple as the one we have described would be elaborate if it were built up out of transistors and other discrete circuit components.

For simplicity, the need for certain control functions in the system has not been taken into account. For one thing, there is obvious need to be able to

Figure 2.1   A pulse counting system.

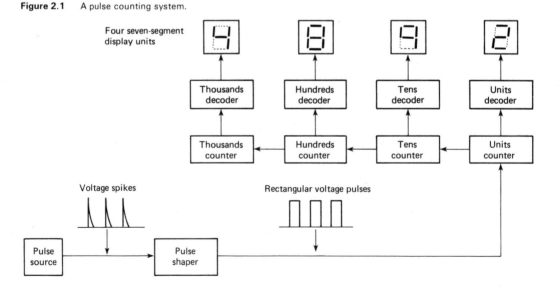

start and stop the counting operation and to be able to *clear* the counters and the display back to 0000 when an entirely new counting run is to be begun. Again, it is desirable to be able to cause the display decoders to *latch* or hold a count in order to read the display panel. If the counting were to proceed at a high rate, the display panel could not otherwise be read. Such expansions of the capabilities of the simple system can also be handled very simply. Sometimes this is done by adding other fundamental integrated circuit building-blocks to the system. Sometimes it is even more easily done since the manufacturers of ICs have already incorporated the extra features needed into their counters, decoders, and other devices.

Analog Versus Digital

The system which we have just discussed is an instance of a digital system. The pulses which are counted are discrete electrical events. At any particular time, the number of past pulses is an integral number such as 8, 9, or 10. As the counting continues, the readout display panel always indicates an integral number.

A digital clock is another example of a digital system. Although time itself goes on increasing *continuously,* the clock reading at any moment can only be to the nearest minute or second, depending on that particular clock.

In a digital system the variable itself (such as the number of automobiles passing a point) inherently comes in "chunks," or discrete steps, or else (as in the case of the digital clock) an inherently continuous variable (time) is broken into discrete chunks before readout.

Now consider, for contrast, a conventional clock with an hour hand only. The hand will move continuously around the face of the dial. You can read off the time to any fraction of an hour in principle, though in practice there will be a limitation to the accuracy with which this can be done. The indication of the time by the hour hand is not presented in discrete steps but in continuous fashion around the dial. This is an example of an *analog* display.

As another example of an analog system, consider the simple circuit of Figure 2.2. As the variable resistance R is changed by turning the potentio-

Figure 2.2   An analog circuit.

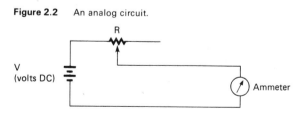

meter knob, the current through the ammeter will vary continuously and the position of the indicater needle on the face of the ammeter will move continuously.

Counting individual objects such as coins or voltage pulses is a digital operation. Measuring something like the height of a growing child is an analog operation.

In this book we will deal only with digital electronics. Analog electronics is a separate subject which is huge and highly divided. It includes audio sound systems and the video information in television transmission and reception. However, as we shall see, what can be accomplished with digital systems is by no means a narrowly limited matter. Digital systems are starting to include audio systems (digital tape recording) and may eventually include TV transmission and reception.[1] It certainly includes anything which can be reduced to the terms of arithmetic or logical functions and that is a great deal indeed!

The NAND Function

In this section we will discuss one of the most fundamental digital concepts— the NAND function.

Figure 2.3 shows a simple circuit which responds to *information inputs* (the status of the two switches) by producing an *information output* (the state of the lamp) as shown in the table of part (b). If either or both of the switches are open, the lamp will be ON. If both switches are closed, the lamp will be OFF, the reason being that a short circuit exists across the lamp. The system gives a "lamp ON" condition if we have *NOT* both A *AND* B closed. Contraction of the words NOT and AND gives rise to the single word NAND.

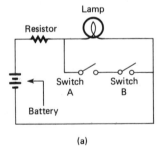

Switch A	Switch B	State of Lamp
Open	Open	On
Open	Closed	On
Closed	Open	On
Closed	Closed	Off

(a)   (b)

Figure 2.3   (a) A NAND circuit; (b) tabular description of the functioning of the circuit in (a).

[1]Excellent pictures of planets and other celestial objects are currently being transmitted back to earth from space vehicles entirely in the form of digital information.

In some circumstances, there might be a door in which there are two locks and which would be considered insecure unless both locks were locked. The system gives a "door insecure" condition if we have NOT both lock A AND lock B locked. From a formal point of view, the descriptions of the circuit and of the door are the same. Each is said to obey the NAND function.

Still other physical systems which conform to the NAND function can be found. It is desirable to have a universal way to represent the logical nature of any such system independently of the details of the system such as switches and locks. Let us recognize that both of the systems we have discussed have two *inputs* (switch states or lock states), each of which can be in one or the other of two conditions (open or closed, fastened or unfastened), and that each system has an output (lamp state or door state) which can be in one or the other of two conditions (on or off lamp, secure or insecure door). We achieve universality by using the symbols 0 and 1 to represent the conditions of each of the inputs and of the output. (Symbols other than 0 and 1 could be used, such as "red" and "green," but the symbols 0 and 1 are ideally suited to interpretation as the digits 0 and 1 of the binary number system, as will be seen, and this constitutes an enormous advantage in the symbolism used.)

Table 2.1 The NAND truth table.

Input A	Input B	Output
0	0	1
0	1	1
1	0	1
1	1	0

Table 2.1 shows in this universal notation what the essential nature of the NAND function is. Whatever the physical system may be, the output is 1 only if it is *NOT* the case that A *AND* B are both 1.

We must pause here to note two matters of terminology. One is that the table in Table 2-1 and others which we will meet when we come to the functions OR, AND, OR, NOR, and a few more are called *truth tables*. The origin of this name lies in the history of the development of the subject of formal logic. In that subject, similar tables arise in connection with the relations between logical premises and conclusions and their truth or falsity.

The other is that in digital integrated circuit electronics, devices which conform to the NAND truth table (and to the tables for the OR, AND, OR, NOR, and other functions) are called *logic gates* or simply *gates*. Why the word "gate" is appropriate will be explained a bit further on. However, we will use such terminology as NAND gate from now on.

The NAND Gate in Integrated Circuit Form

Consider a certain circuit which has been constructed by someone from transistors and other components. The circuit, once built, has been packaged in a closed chassis so that the actual circuitry is not visible but has five accessible wire leads issuing from the box. The gadget appearance is shown in Figure 2.4. In the figure, the leads a through e have been connected to external switches, two voltage sources, and three voltmeters.

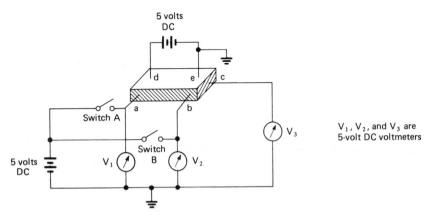

Figure 2.4 A black-box NAND circuit.

Now suppose that we were to experiment with this "black-box" circuit. In doing so, we could manipulate switches A and B and observe what the voltmeters indicate in response. Since leads d and e are permanently connected to one of the power supplies (in order to supply power to the internal transistors), we have only the two inputs, a and b, to manipulate and one output, c, to watch. We can set the states of the switches and observe the resulting voltmeter readings and do nothing more. Suppose finally that the results of the experiments are as shown in Table 2.2.

Table 2.2 Results of experiments with the circuit of Figure 2.4.

Switch A	Switch B	Lead c Voltmeter Reading	Lead a Voltage Level	Lead b Voltage Level	Lead c Voltage Level
Open	Open	+5V	0V	0V	+5V
Open	Closed	+5V	0V	+5V	+5V
Closed	Open	+5V	+5V	0V	+5V
Closed	Closed	0V	+5V	+5V	0V
	(a)			(b)	

If in part (b) of Table 2.2 we use the binary digit 0 to indicate zero voltage level for any wire lead and the binary digit 1 to indicate a +5-volt level for any wire lead, then the behavior is as shown in Table 2.3(a). Since this table is identical with Table 2.1, the gadget is in fact a NAND gate.

Table 2.3 Other ways to represent the logic of Table 2.2.

V at a	V at b	V at c	Logic level at a	Logic level at b	Logic level at c
0	0	1	Low	Low	High
0	1	1	Low	High	High
1	0	1	High	Low	High
1	1	0	High	High	Low
	(a)			(b)	

In part (b) of Table 2.3 we show the same truth table as in part (a) but we use a very common alternate mode of expression. *Logic low* or simply *low* is equivalent to binary 0 and *logic high* or *high* is equivalent to binary 1.

The black-box device shown in Figure 2.4 has been introduced in this section because of its extremely close resemblance to a modern integrated circuit NAND gate. The device is a closed package which contains transistor circuitry. To put the device to practical use[2] you ordinarily are not concerned with the internal circuitry but only with the fact that if it is powered properly at leads d and e, then it will respond to inputs at a and b by producing outputs at c according to the nature of a NAND gate.

Similarly, the modern integrated circuit form of the NAND gate is a package with sealed-in internal circuitry and only its external pins (leads) accessible. Two-input NAND gates (which is what we have been discussing) come four to a package, called simply a *quad* package. Figure 2.5 shows

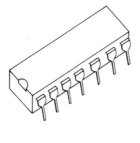

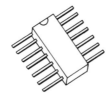

Figure 2.5   A DIP IC package and a flat-pack IC package.

[2]Some applications for NAND gates will be discussed in the next chapter.

what a quad NAND IC looks like. This is a 14-pin *DIP (Dual In-line Package)* format. Of the 14 pins, two are for power supply connections and the other 12 are allocated to the four NAND gates, two as input pins and one as an output pin for each gate. A glance ahead to Figure 2.7 will show you a representation of the internal arrangement of the gates (each represented by ⊐Dᴏ) and the allocation of the pins.

More About IC Packages

Figure 2.5 shows, along with the DIP integrated circuit device, another format known as a *flat-pack*. The flat-pack format is used by manufacturers who must utilize low-profile packages to conserve space. DIPs are in almost universal use otherwise. They are especially easy to work with because auxiliary apparatus is readily available. Among the accessories available are sockets into which the ICs can be inserted and DIP *breadboards* (boards for constructing temporary, easily modifiable circuits for experimentation).

In the rest of this book we will have only DIPs in mind. However, the logical concepts and most of the practical matters we will discuss apply equally well to DIPs and flat-packs.

In addition to the 14-pin DIPs as shown in Figure 2.5, there are 16-pin and larger (24-pin and 48-pin) packages. Figure 2.6 provides more detailed pictures of the 14-pin package. These are about 1.5 times life-size. The pin numbers shown in part (a) are not actually printed on the body of the chip. Various conventions are used to label the pins in circuit diagrams. We show pin numbers and the names of pin functions inside the package outlines except where clarity is advanced by some other usage.

It is essential that you be able to identify the pins of an IC device. The pins are numbered in counterclockwise fashion starting with pin 1 at the lower left corner of the package as seen from the *top* of the package. The left end is indicated by a mark, sometimes a circular indentation near the edge and sometimes a semicircular depression at the edge. The nature of the mark and the sequential numbering of the pins is shown in Figure 2.6.

Every IC must be supplied power in order for its internal circuitry to function. This requires one pin for connection to the power supply "high"

Figure 2.6   More detailed views of a DIP IC.

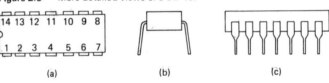

(a)   (b)   (c)

terminal and one pin for connection to the power supply "low" terminal. (More will be said about power supply requirements in due course.) Most often, the corresponding power supply pin numbers are 14 and 7 respectively for 14-pin packages and 16 and 8 for 16-pin packages. However, these are not universal rules and the only safe procedure is to refer to the specifications of any IC which you intend to use.

You also need to know the functions of all the other pins of your IC. To illustrate a succinct way in which this information is often given, we present Figure 2.7. Such a diagram is called a *pin diagram,* appropriately enough. Such a diagram gives only bare bones information but such information is often all you need. When more detailed information is needed, such as the amount of load the device can drive or the maximum speed of operation possible, you must refer to more complete technical descriptions of the device you are using.

In the case of the pin diagram of the quad NAND gate device shown in Figure 2.7, the symbol ⊐D● used for each of the four NANDs tells us what the logical function is. Each NAND obeys the truth table shown in Table 2.3. If the IC were more sophisticated or less familiar than the well-known NAND gate, its pin diagram would have to be accompanied by a truth table as the pin diagram alone would not be able to make the logical function of the device clear.

The quad two-input NAND gate shown in Figure 2.7 is also identified by a number, 7400. You never find the name of an IC device printed on it. They are all identified by number. The 7400 is one of an extensive family of devices known as the *7400 series* collectively. We will presently meet other instances

Figure 2.7   Pin diagram of the 7400 quad NAND gate IC.

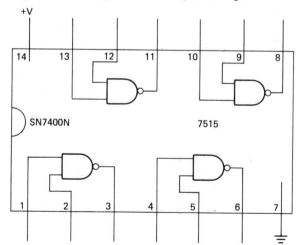

such as the 7404, 7448, and 7490. (These are not NAND gates.) The 7400-series devices are *TTL devices.* TTL devices make up a large *logic family* which is one of the two most popular families in general use today. The other popular logic family is composed of CMOS devices. The meanings of the terms TTL and CMOS and technical facts about them will be discussed later in the book. In particular, in Chapter 5 we will discuss some special properties of TTL devices. It will be pointed out there that standard 7400-series TTL ICs in some respects have technical characteristics which may not be adequate for specialized applications. There is a series of TTL devices which have improved characteristics. These are logically the same as their 7400-series counterparts but they belong to a 5400 series. Thus a 5400 two-input NAND gate is logically the same as the 7400 but (for one thing) it can tolerate a wider range of ambient temperatures, making it better for applications in aircraft and space vehicles. Laboratory scientists, experimenters, and hobbyists do not often encounter 5400-series ICs.

In Figure 2.7 there are some letters and numbers on the pin diagram to illustrate the kind of printed markings you typically find on the upper face of an IC. In this example, "7400" identifies the nature of the device. "SN" identifies the manufacturer as the Texas Instruments Corporation. "7515" means the fifteenth week of the year 1975, and is when the device was made. (In some circumstances, usually industrial, it may be important to be able to date the manufacturing run which produced a device.) The "N" following "7400" means "dual in-line *plastic* package." There are also ceramic packages which are hermetically sealed more thoroughly than are plastic packages and one of these would be labelled with "J."

Here is another example. "NS DM7400J 628" on an IC would indicate that the National Semiconductor Corporation made that particular 7400 device in the 28th week of 1976 and it is in a ceramic package. "DM" is a marking peculiar to the company.

The 74XX numbering system has become standardized so that throughout most of the industry a number such as 7400 has the same meaning. There are, however, exceptions. The Motorola Corporation calls their equivalent of the 7400 an 0905. Other companies using different numbering systems are the Fairchild Camera and Instrument Corporation which has an equivalent 9000 series and Signetics with an equivalent 8000 series.

More About TTL and
Other Families of Integrated Circuits

In the preceding discussions we have referred to some devices as being TTL. The 7400 series which includes the 7400 NAND consists entirely of TTL devices. In the earlier days of the IC era, devices were produced which used circuitry where interconnected diodes were prominent. These came to be known as *DDL* or *diode-diode logic* devices. *DTL* devices also appeared, this term meaning *diode-transistor logic*. Even more types existed such as *RTL*, resistor-transistor logic. These older families of ICs are relatively little used today except in special circumstances but many of them are available from electronics dealers.

TTL integrated circuits are made predominantly from transistors. The name means transistor-transistor logic. You may know that two kinds of transistors of the bipolar type exist, the NPN and PNP, and that these can be fabricated of silicon or germanium. For technological reasons, modern TTL integrated circuit devices use silicon NPN transistors.

These TTL devices are very admirable. They are excellent for high-speed digital circuits, readily available, inexpensive, reliable, and easy to work with. The TTL family is not perfect, however. One drawback in many applications, as in the case of battery-operated systems, is its relatively high current drain from the power supply.

At the present time there is one principal competing family of integrated circuit devices. It is known as *CMOS*, the abbreviation standing for *complementary metal oxide semiconductor*. In CMOS ICs, the active components are certain monopolar transistors known as MOS devices. The chief desirable feature of CMOS chips is that they draw virtually no current from the power supply when they are in the standby mode and very little even when they are in operation. CMOS integrated circuit devices will be discussed in Chapter 15 as will older families and more recent families which include *ECL* (Emitter-Coupled Logic) and *IIL* or *I²L* (integrated-injection logic).

SSI, MSI, LSI, and Other Scales of Integration

Digital integrated circuit devices are made in various degrees of internal complexity. A single NAND gate, which might be any one of the four in a 7400 TTL chip, is made from an arrangement of four NPN transistors and a diode. A 7400 therefore contains 16 NPN transistors and 4 diodes. This is said to be a case of small scale integration or *SSI* for short.

In a four-bit BCD counter (which we will discuss later) the number of transistors may be in the neighborhood of 250. This is regarded as medium scale integration or *MSI*. Some simple memory devices also fall into the MSI class.

A certain memory device is known as a 1-K device, which means it can store approximately 1000 bits (0's and 1's) of information. It may contain around 4000 transistors. This is a case of large scale integration or *LSI*.

During the years from about 1960 to the present, the number of components packaged into IC devices has increased at a surprisingly uniform rate.[3] The trend is continuing and is leading to the production of devices in which the number of internal components will be staggering, exceeding (currently) 200,000. These are referred to as very large scale integration devices or *VLSI*. Ultralarge scale integration or ULSI is already appearing over the horizon. However, VLSI and ULSI are possible only because these densely packed devices use new physical principles and not those of the relatively primitive (already!) NPN transistors or MOS components of the TTL and CMOS families.

EXPERIMENTAL WORK—HARDWARE

At the ends of most of the chapters in this book, beginning here, we will give suggestions for building and studying circuits. At this point Appendix C should be read. This appendix deals with the hardware that is needed for IC experimentation.

A first IC experiment

This chapter has discussed NAND gates. Experiments with single NAND gates, as opposed to experiments with systems in which several NANDs are used, must of necessity amount to no more than verifying that a NAND gate behaves as it should. However, an experiment of that kind at this early stage in our study of ICs can be valuable so you can get familiar with breadboards and other apparatus, and to develop facility in wiring up circuits quickly and without error.

You should have read Appendix C and, in particular, should understand the Solderless Breadboard section of that appendix. This section discusses the essentials of wiring up a circuit on a breadboard.

Figure 2.8 shows the circuit to be built. The NAND gate is one of the four in a 7400 quad NAND device. (Any of the four can be used.) The pin 7 and pin 14 connections are to supply power for the 7400. Switches A and B are to make it possible to set the logic states of the inputs to the NAND gate at logic 0 or logic 1 manually. In order to observe what is the logic state of the output pin 3, the circuit uses any common LED (light-emitting diode) with a current-limiting resistor in series with it. (More about this can be found in the Logic State Indicator section of Chapter 9.) Alternatively, a DC voltmeter capable of reading +5 volts can be used in place of the LED and the resistor.

---

[3]A readable and instructive article on the topic is "Microelectronics" by Robert N. Noyce, *Scientific American*, September 1977, pages, 63–69.

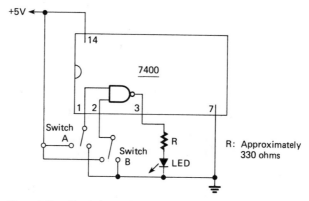

Figure 2.8   Circuit for verifying the functioning
of a single NAND gate.

As an example of what to expect, consider the case in which switch A is connected to the +5-volt source and switch B is connected to the circuit ground. This corresponds to inputs 1 and 0 to the NAND and the output should be 1. The light should go on, or the voltmeter should indicate logic high. In similar fashion the rest of the NAND truth table should be verified.

3

NANDs Again—
Some Uses and Varieties

The first part of this chapter is intended to give you an insight into the way in which simple logic gates can be combined to produce a *system* which can perform functions at a higher level of complexity than can the gates themselves. Until now we have discussed only the logical nature of the NAND gate. In Chapter 4 we will go on to discuss the other gates (OR, NOR, and so on). However, it is instructive to see what can be accomplished with just one kind of gate. In doing this, we will introduce some related ideas.

Many systems built from the basic logic gates carry out what is called *combinatorial logic*. Given only the logic states of all the inputs to the system, the structure of the system itself results in unique states for all outputs, which do not depend on anything the system did previously. In another kind of system, what the outputs are depends not only on the current input states, but also on what the input states were earlier before they were changed to their new values. This is a distinctly different kind of system and it is said to carry out *sequential logic*. Sequential logic will begin to make its appearance incidentally in some later chapters and will be dealt with squarely in Chapter 10. In this chapter, we will concentrate on combinatorial logic.

Finally, we will see that NAND gates with more than two inputs exist and we will describe several popular TTL NAND ICs.

The Half-Adder

An important circuit is one called the *half-adder*. This is a circuit which will accept two one-digit inputs and produce as output the sum[1] of the input digits. This will be a combinatorial logic circuit constructed entirely of two-input NAND gates.

Before going further, let us agree to follow the universal practice of replacing the long phrase "one-digit binary number" by the much shorter word "bit." A *bit* is simply a 0 or a 1.

When two bits are added together there are four possible cases. These are $0 + 0 = 0$, $0 + 1 = 1$, $1 + 0 = 1$, and $1 + 1 = 10$. In decimal notation these say that $0 + 0 = 0$, $0 + 1 = 1$, $1 + 0 = 1$, and $1 + 1 = 2$. Notice that in $1 + 1 = 10$, expression of the sum requires the use of two bits. For uniformity all of the sums could be written using two bits so that they would be 00, 01, 01, and 10, respectively. Indeed this is quite necessary when the addition of two one-bit numbers is part of the addition of two longer numbers such as 11011 and 10101. (See the Addition of Two Binary Numbers section of Appendix A.) When each sum is written as a two-bit number, the rightmost bit is called the *sum* bit, and the leftmost bit of the two is called the *carry* bit. This is shown in full in Table 3.1.

Table 3.1 All of the possibilities when two one-bit numbers are added together.

Bit A	0	0	1	1
Bit B	0	1	0	1
Sum bit	0	1	1	0
Carry bit	0	0	0	1

Let us use these ideas and terminology to summarize what the half-adder circuit is to do: It is to accept two one-bit inputs (A and B) and produce in each of the possible cases two one-bit outputs (sum and carry) according to the truth table of Table 3.1.

The circuit shown in Figure 3.1 will accomplish the purpose. To verify that it does indeed do so, you should assume as inputs A and B the various states which are possible and in each case verify that the sum and carry outputs are correct. We give only one of the four analyses as an example. In Figure 3-2, the inputs A and B are assumed to be each 0. All of the logic states at the inputs

---

[1]If you are unfamiliar with the binary number system you should digest Appendices A and B now.

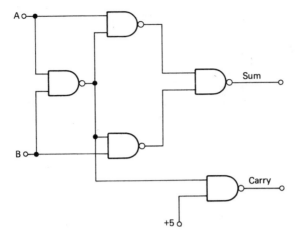

Figure 3.1   A half-adder.

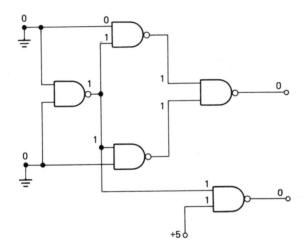

Figure 3.2   Analysis of the half-adder in a particular case.

and outputs of the NAND gates are shown explicitly. Indeed, the result is "sum = 0" and "carry = 0." The other three cases are left to you as an exercise.

Notice that the truth table for this circuit is a two-input, two-output truth table. It is of a higher order of complexity than is the two-input, one-output truth table of the bare NAND.

Wiring Up the Half-Adder Circuit

We wish to show how the half-added circuit of Figure 3.1 can actually be constructed using 7400 ICs and other hardware. A half-adder in a permanently wired digital electronic system would, of course, be solidly soldered into a

20

printed circuit board or the equivalent. However, what we want here is a temporary experimental set-up. The binary digits to be added (A and B) are to be determined by manually operated switches. The states of the outputs are to be indicated by some kind of indicators such as lamps. If a lamp is on, we will read that output state as "1" and if the lamp is off, we will read that output state as "0."

The circuit is shown in Figure 3.3. A convenient basic piece of hardware for this kind of breadboarding of an experimental circuit is a multiple-socket structure into which you plug the ICs and the ends of the leads on connecting wires, resistors, or other circuit elements. This kind of breadboard is discussed in Appendix C where we also describe the power supplies and other apparatus you need for experimenting with ICs.

It is assumed that the input states can be distinguished by the position of the switches A and B. Some kind of indicating devices are needed at the outputs so that one can tell what the "sum" and "carry" states are. (Such indicators *could* also be used at the inputs.) The indicators could be any DC voltmeters with full-range capabilities sufficient for reading maximum voltages of about +5 volts. They could, with even greater convenience, be light emitting diodes (LED). A LED is a solid state lamp which emits light when its anode lead is raised to a suitable positive voltage above the circuit ground and its cathode lead is connected to the circuit ground, but which is off when the anode lead goes to a voltage low state. LEDs are discussed in more detail at the beginning of Chapter 9.

Figure 3.3   Actual construction of the half-adder circuit.

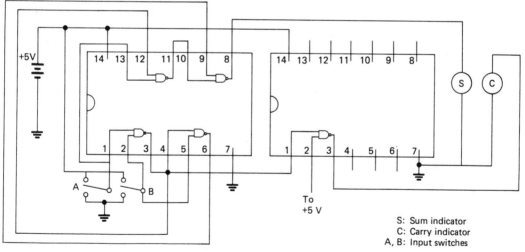

S: Sum indicator
C: Carry indicator
A, B: Input switches

The circuit of Figure 3.3 may seem to the inexperienced reader to be somewhat complex. However, it is easy for even the rawest beginner to wire up and it makes a fine first laboratory exercise. Once the circuit has been built, it is a simple matter to operate the switches, watch the sum and carry indicators as they respond to the throws of the switches, and verify that the circuit really functions as a half-adder. We urge the reader (whether you build the circuit or not) to examine the circuit diagram and satisfy yourself that it really represents a wired-up form of the more abstract diagram shown in Figure 3.1.

An Important Special Property of TTL ICs

Consider the circuit shown in Figure 3.4. The ground pin (7) and the +5-volts pin (14) have been connected to a suitable external power supply. Another pin is assumed to be any input pin for the device and it is connected to a switch that is wired into the circuit so that the state of the input pin can be made high or low at will. It appears, with respect to the input pin, that there are precisely the following possibilities: If the switch is thrown to connect the pin to the circuit ground, the logic state of the pin is 0 and, if the switch is thrown to connect the pin to the +5-volt terminal of the power supply the logic state of the pin is 1. Actually, there is a third way to treat the input pin. If the input pin is not connected either to the circuit ground or to the +5-volt terminal, the pin acts as though it *were* connected to the +5-volt terminal. This is stressed in Figure 3.5 where the single-pole, single-throw switch allows only connecting the input pin to the circuit ground or leaving the pin floating. In many cases, the circuit of Figure 3.5 will serve as well as the circuit of Figure 3.4. In brief, a floating TTL input pin can act as if it were at logic high.

This rule often helps to simplify labbench circuits for simple experimental work by eliminating the need for switches and wire leads which contribute to clutter. You can simply attach one end of a wire to an input pin and push the other end into a circuit ground socket of the breadboard or pull that end out of the socket, in order to make the pin low or high. However, this rule comes accompanied by three strong warnings. One is that it does not always

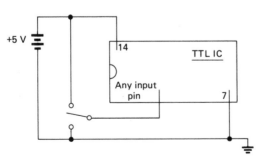

Figure 3.4   A circuit for setting the logic states of an IC input pin.

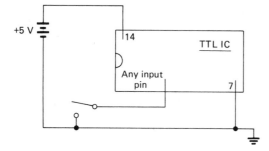

Figure 3.5   A sometimes acceptable
alternative to the circuit of Figure 3-4.

work. Another is that even if it works in a labbench experimental circuit, it is very bad practice to leave TTL input pins floating in hard-wired permanent instrumentation because an unconnected input pin is suceptible to pick-up of electrical noise. The third is that the rule applies only to TTL integrated circuit devices and, in particular, not to ICs of the CMOS family.

You must also know that mechanical switches as might be used in the circuit of Figure 3.5 suffer from a problem called "bounce." *Bounce* means that on make or break, a switch does not cleanly make once or break once but in effect closes or opens several times before finally settling down to a definite condition. This is equally true if you push in and pull out a wire end from a socket in a breadboard to avoid using a switch. (Switch bounce and its cure are discussed further in Chapter 4.)

The Full-Adder

A *full-adder* circuit is a circuit which can accept as inputs a bit A, a bit B, and a carry bit C from some previous addition and produce as outputs the proper sum bit and the proper carry bit. It is therefore a three-bit input, two-bit output circuit.

To illustrate how the need for a full-adder can arise, you might consider the problem of adding the binary numbers 01 and 11 together. In decimal notation, the problem is to add 1 (decimal) and 3 (decimal) together. The sum is to be 4 (decimal).

The problem and its solution in binary notation reads as follows: $01 + 11 = 100$. How the addition of 01 and 11 is carried out stage by stage is shown in Table 3.2.

Table 3.2   Step-by-step addition of two two-bit numbers.

						Carry from right column addition			
				1				1	
0	1	0	1	0	1			0	1
1	1	1	1	1	1			1	1
		0 Right column sum		0	Left column sum		1	0	0 The sum
		1 Right column carry		1	Left column carry				
(a)		(b)		(c)				(d)	

In part (a) the problem is restated. In part (b) the rightmost bits of the two-bit addends are added. The sum requires a sum bit and a carry bit for its expression. In part (c) the carry bit from the rightmost column addition is added to the other bits in the leftmost column. Then the sum required in part (a) can be seen to consist of *three* bits, as shown in part (d) of the table. However, the addition in any one column is always a matter of adding together three one-bit numbers and producing a suitable sum bit and a suitable carry bit.

In Figure 3.6 we show a way in which two of the half-adders of Figure 3.1 can be combined with some NAND gates to produce a full-adder. You should analyze the functioning of this circuit in response to the various possible inputs A, B, and C to see whether the circuit shown is really a full-adder or not.

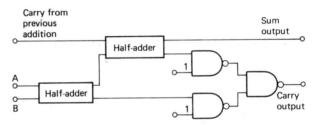

Figure 3.6 A full-adder constructed from half-adders and logic gates.

The Importance of the Full-Adder Circuit

If enough full-adders are combined into a single system, the system can be made to add a binary number of considerable length to another binary number of comparable length. Thus, two 32-digit binary numbers could be *added* together. Since *multiplication* can be regarded as successive additions, it would be possible to carry out the multiplication of two binary numbers of arbitrary length by using IC adders. One way in which a binary number can be *subtracted* from another binary number is to form the *two's complement*[2] of the first number and then to add together that two's complement and the second number. It is not difficult to use digital IC devices to form the two's complement of an arbitrary binary number and we already have a way to add binary numbers. It follows that with adder circuits and a bit of other IC circuitry, subtraction of binary numbers can be carried out. Since *division* can be re-

---

[2]The *two's complement* of a binary number is defined and discussed in Appendix A. It is not necessary to know this material in order to understand the point being made here.

garded as successive subtractions, it follows that with adders and whatever is required for forming two's complements, division becomes possible.

In this way you can see how, in principle, all of the four arithmetic operations $(+, -, \times, \div)$ can be carried out using digital ICs. But then, in principle, it becomes possible to do an enormous amount of the world's calculational work. Keeping bank accounts, analyzing the national census data, and just pushing the buttons on a simple pocket "four-banger" calculator to compute a simple arithmetic result—all this and much more can be accomplished.

In this discussion, the phrase "in principle" has appeared repeatedly. The reason this somewhat cautionary phrase has been used is to avoid suggesting that the scheme outlined is the only, or even the most efficient, scheme for carrying out the program. Actually, there are simpler and much more efficient ways to do arithmetic. Nevertheless, we have thought it an impressive way to hint at the enormous possibilities which begin to open up when we have gone only a little way into the subject of digital ICs.

Some Higher Order NAND Gates

Until now we have restricted our discussions of NAND gates to a type which has two inputs and one output. It is possible by extension to conceive of a system which behaves in a similar way but which has more than two inputs and yet only one output. As an example we present in Table 3.3 the truth table for a three-input NAND gate. NANDs with four and more inputs also exist.

Table 3.3 Truth table for a three-input NAND.

Inputs			Output
A	B	C	Q
0	0	0	1
0	0	1	1
0	1	0	1
1	0	0	1
1	1	0	1
1	0	1	1
0	1	1	1
1	1	1	0

There is a commonly used convention used to allow a NAND gate to be easily represented in a circuit diagram. Figure 3.7 (a) shows the symbol for a two-input NAND and Figure 3.7(b) shows a four-input NAND.

(a)   (b)

Figure 3.7   Symbols for (a) a two-input NAND; and (b) for a four-input NAND.

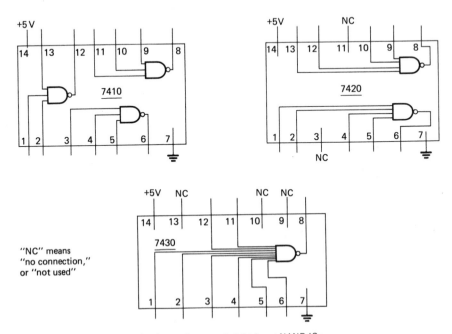

"NC" means
"no connection,"
or "not used"

Figure 3.8   Pin diagrams for three-, four-, and eight-input NAND ICs.

Some Popular TTL NAND Packages

To conclude this chapter, we describe some TTL NAND packages other than the 7400 quad two-input NAND, which has already been discussed in some detail. The three-, four, and eight-input NANDs illustrated in Figure 3.8 are selections we have made on the grounds of the likelihood of their usefulness for beginning experimenters and on the grounds of their general availability.

EXPERIMENTS

1. The half-adder

The idea in this experiment is to make a half-added using NAND gates and to verify that the circuit performs the functions for which it is intended. Figure 3.1

shows the logic of the circuit and Figure 3.3 shows the circuit in detail, as it could actually be constructed.

2. The full-adder

A more elaborate project would be to make a full-adder circuit. For this, the circuit of Figure 3.6 can be referred to. That we do not give a highly detailed circuit drawing, including (for example) specific pin numbers of specific TTL devices, in Figure 3.6 is deliberate. You very much owe it to yourself to begin early on to start an experimental project by making your own circuit drawings, then building the circuit from your own drawing. This enhances both your understanding and your practical abilities, and leads to a sense of growing mastery.

3. Design and construction of a circuit

A fine next step after making a detailed circuit drawing from a schematic representation of only the logic of the circuit as in Experiment 2 is to carry out the complete design of a circuit, starting only with the truth table that defines what the circuit is to do. We suggest that you tackle the design of a circuit which satisfies the truth table shown in Table 3.4. Then the circuit should be built and tested for proper functioning. Incidentally, this circuit is one known as an OR circuit. The OR function will be discussed fully in the next chapter.

Table 3.4 Truth table for Experiment 3.

Input A	Input B	Output Q
0	0	0
0	1	1
1	0	1
1	1	1

The Other Logic Gates

The AND Gate

The purpose of the experimental set-up shown in Figure 4.1(a) is to illustrate how the need for a logical function other than the NAND function can arise. The radioactive sample is a photon emitter. It spews out particles of electromagnetic radiation which are called *photons*. Sometimes two photons will be emitted simultaneously in opposite directions from the source. Sometimes two photons will be emitted simultaneously but with some nonzero angle A between the lines of emission. The purpose of the experiment is to collect data which shows the rate per unit time with which photons from the source arrive simultaneously at the detectors (within the resolution limits of the apparatus) as a function of the angle A between the detectors. We assume that the detectors send voltage pulses of suitable size and shape to the "coincidence unit" and that the latter obeys the truth table shown in Figure 4.1(b). In that table any 0 means "no count" and any 1 means "count."

The behavior of the coincidence unit can be summarized in this statement: The output is 1 if there is 1 at input A *AND* 1 at input B simultaneously, otherwise the output is 0.

You can think of other systems which exhibit the same logical behavior. Figure 4.2 shows one such system. The lamp will light only if switch A *AND* switch B are closed.

Any system which behaves in the same manner is called a two-input AND gate and is represented by the symbol shown in Figure 4.3. By extension of the idea of the two-input AND we could have a device of the kind shown in

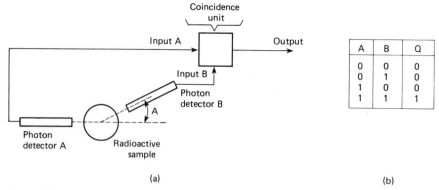

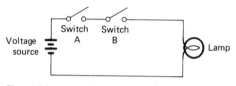

A	B	Q
0	0	0
0	1	0
1	0	0
1	1	1

(a)   (b)

Figure 4.1   (a) An experiment in which the AND function is needed; and (b) a truth
table for (a).

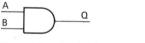

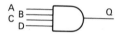

Figure 4.2   An AND circuit.

Figure 4.3   Symbol for a two-input
AND gate.

Figure 4.4   Symbol for a four-input
AND gate.

Figure 4.4. In this four-input AND, the output is a 1 only if input A *AND* input
B *AND* input C *AND* input D are 1's.

The pin diagrams for some of the most commonly used TTL integrated
circuit ANDs are shown in Figures 4.5, 4.6, and 4.7. These are a quad two-
input AND, a triple three-input AND, and a dual four-input AND, respectively.

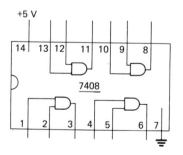

Figure 4.5   Pin diagram of the
7408 quad two-input AND IC.

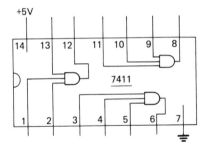

Figure 4.6   Pin diagram of the 7411 triple three-input AND IC.

"NC" means "no connection" or "not used"

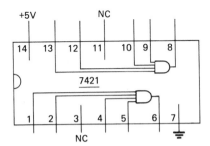

Figure 4.7   Pin diagram of the 7421 dual four-input AND IC.

The Inverter and the Buffer

The *inverter,* also called a *NOT,* has the truth table shown in Figure 4.8(a). The function performed by this device is simply to turn an input 0 into an output 1 and to turn an input 1 into an output 0. The inverter is said to *complement* the input. This is a one-input, one-output function and so it is different in nature from an AND or a NAND. However, it has become customary to class the device among the logic gates with ANDs and NANDs.

The conventional symbol for an inverter is shown in Figure. 4.8. The main triangular part of the symbol has its origin in electronics, where such a triangle represents an amplifier. The small circle at the right vertex of the triangle explicitly indicates that the output of the device is the complement of the input. It is standard notation today to indicate the complement of any quantity (such as the input A of the inverter) by putting a bar over the symbol that represents the quantity. Hence the inverter has output $\bar{A}$ if its input is A.

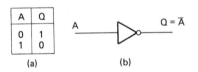

Figure 4.8   (a) Truth table for an inverter; and (b) symbol for an inverter.

A	Q
0	1
1	0

30   (a)   (b)

These observations make it possible to explain a connection between the symbols for the AND gate and the NAND gate. For convenience, the corresponding truth tables are reproduced in Table 4.1. You can see that an AND with its output followed by an INVERT is a system which performs a NAND function. How this might be done in practice is shown in Figure 4.9. Similarly a NAND and an INVERT can be combined to make, in effect, an AND.

Table 4.1   Truth tables for an AND and for a NAND compared.

A	B	Q	A	B	Q
0	0	0	0	0	1
0	1	0	0	1	1
1	0	0	1	0	1
1	1	1	1	1	0
	AND			NAND	

Figure 4.9   An AND and an inverter combined to form a NAND.

Figure 4.10   Symbol for the two-input NAND.

The symbol in Figure 4.10 is one we have met before. It is the symbol for an AND with an INVERT at the output point, making it the symbol for a NAND.

The workhorse inverter IC is the 7404 hex inverter. *Hex* means that the device contains six individual inverters. The 7404 pin diagram is shown in Figure 4.11.

Figure 4.11   Pin diagram of the 7404 hex inverter IC.

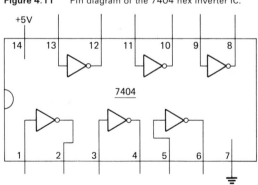

A device closely related to the inverter is one known as a *noninverting buffer*. It is used between a signal source and a load in order to provide load-driving capability greater than that which the signal source itself could give. The truth table is shown in Table 4.2. While it is logically uninteresting, it may be electrically useful. A typical buffer is the 7407 hex noninverting buffer. The output can withstand 30 volts and *sink* 30 milliamperes of current.[1] These specifications exceed those for ordinary logic gate outputs. The appearance of 7407 is shown in Figure 4.12.

Table 4.2 Truth table for the noninverting buffer.

A	Q
0	0
1	1

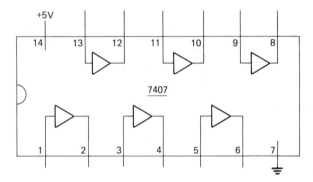

Figure 4.12   Pin diagram of the 7407 hex noninverting buffer IC.

OR, NOR, Exclusive OR, and Exclusive NOR

The list of the basic logic gates will now be completed. The natures of the two-input OR, NOR, XOR (exclusive OR), and XNOR (exclusive NOR) gates are shown in the truth tables of Figure 4.13(a). The corresponding conventional circuit symbols for the gates are shown in Figure 4.13(b).

It is not hard to fix in your mind what the function is in each of these four cases. For example, you can remember that the OR gate gives an output 1 if either A *OR* B or both is 1, as the name OR suggests, and that NOR gives an input 1 if *Neither* A *OR* B is 1, as the name NOR suggests. Pin diagrams of commonly used 7400-series IC gates of each of the kinds listed above are shown in Figure 4.14. (The 74266 is an open collector device.)

---

[1]"Current sinking" and "current sourcing" are matters discussed more fully in Chapter 5. The 7407 is an "open collector" TTL IC, in which respect it is different than the majority of TTLs. Open collectors ICs are also discussed in that chapter.

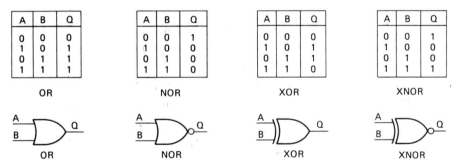

Figure 4.13 Symbols for the OR, NOR, XOR, and XNOR logic gates and their truth tables.

Figure 4.14 Pin diagrams for (a) 7432 quad two-input OR IC, (b) 7402 quad two-input NOR IC, (c) 7486 quad two-input XOR IC, and (d) 74266 quad two-input XNOR (open collector) IC.

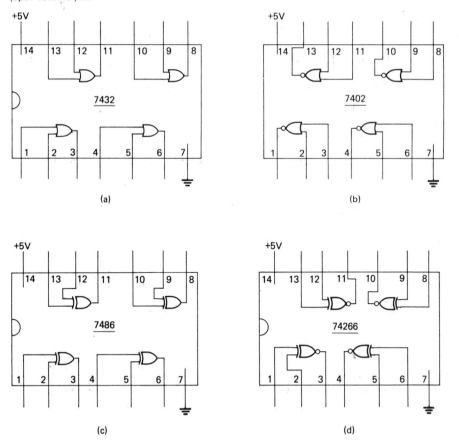

Why Are the Logic Gates Really Gates?

The little circuit shown in Figure 4.15 has an important purpose. Coming in on the line which leads to input B of the AND is a series of 0 and 1 voltage pulses. It is desired that by operating the switch, you can at will determine whether the pulses are passed through to the output Q or not. If the switch is closed, input A of the AND is 0 and the output Q will be 0 *regardless* of whether the input B is 0 or 1. If the switch is opened, input A is 1, and whether the output Q is 0 or 1 depends on whether input B is 0 or 1, respectively. Thus with the switch open, the pulses going to input B are passed through and come out at Q. This resembles an actual gate which, if open, passes people through. If the switch is closed, there is analogy with a closed gate. This is the origin of the word "gate" in the name "AND gate."

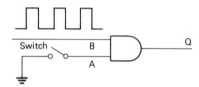

Figure 4.15   The AND as a gate.

The process of operating input A to pass or to block the pulses at input B is called "gating." In the circuit shown, the gating is accomplished by manual operation of the switch but in a more elaborate circuit the state (0 or 1) of input A may be determined automatically by a signal (0 or 1) from some other part of the integrated circuit system.

NAND, OR, NOR, and the other basic logic ICs may be used in similar fashion but then the output Q may not be an exact duplication of the sequence of pulses at input B. For example, if a NAND gate is used and input A is 0, the output Q is 1 regardless of the input signal at B, but if input A is 1, the output Q is the complement of the state of input B, so that 0 at B comes out as 1 and 1 at B comes out as 0. Nevertheless, it is the custom to call these devices gates. They are gates which may *process* (change) what passes through them when they are open.

Some Examples of the Use of the Basic Logic Gates

The two-line to four-line decoder
Consider the truth table given in Table 4.3. This is a two-input, four-output truth table.

Table 4.3 An example of a two-input, four-output truth table.

A	B	Q_a	Q_b	Q_c	Q_d
0	0	1	0	0	0
0	1	0	1	0	0
1	0	0	0	1	0
1	1	0	0	0	1

There are four possible combinations of 0 and 1 at the inputs A and B. In response to any one of the four input states of A and B, one and only one of the four outputs becomes 1, the rest remaining 0. A system which accomplishes what the truth table says transforms input data from two lines into output data on four output lines. You can say that it *decodes* the two-line information into a different form—four-line information. Hence the name "two-line to four-line decoder."

Suppose that the purpose of the system is to light up one of a bank of four indicators, such as LEDs, so that whichever of the four indicators is on will indicate which of the four input combinations is present. You might well call such a scheme an *interpreter* or *one-of-four indicator* as well as a decoder.

What we are interested in at present is how the basic logic gates can be used to make a system which will obey the truth table of Table 4.3. One way is shown in Figure 4.16. You owe it to yourself to verify that the circuit shown does what it is intended to do. This circuit is not the only one which will achieve this end. You should try to think of other solutions to this problem.

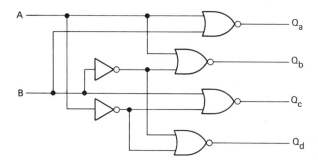

Figure 4.16 A circuit that acts as a two-line to four-line decoder.

Another example of the use of the logic gates

Suppose that in a certain experiment a circuit which has the truth table of Table 4.4 is needed. You can see by inspection that this truth table is not one of the standard set (AND, NAND, OR, NOR, XOR, XNOR, or INVERT). What is· needed to accomplish the job is a combination of the basic logic gates. One way to build the desired circuit is shown in Figure 4.17.

Table 4.4 An arbitrary truth table with two inputs and one output.

A	B	Q
0	0	1
0	1	0
1	0	1
1	1	1

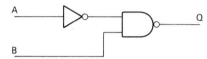

Figure 4.17   A circuit that satisfies the truth table given in Table 4.4.

The truth table for this circuit is still another two-input, one-output truth table. This leads us to an interesting question. How many different two-input, one-output truth tables are there? Since each input A and B can be in one of two states (0 or 1) there are four possible two-input combinations, and since for each of those four input combinations there are two possible output states (0 or 1), there are $2 \times 2 \times 2 \times 2$ or 16 possible truth tables. However, some of the truth tables are very uninteresting. The first two truth tables in Table 4.5 show useless instances.

However, the third and fourth of the truth tables are nontrivial. As a check on your understanding, you should try to design circuits that will behave according to those truth tables using the basic logic gates.

Table 4.5 More of the 16 possible two-input, one-output truth tables.

A	B	Q	A	B	Q	A	B	Q	A	B	Q
0	0	1	0	0	0	0	0	1	0	0	1
0	1	1	0	1	0	0	1	0	0	1	1
1	0	1	1	0	0	1	0	1	1	0	0
1	1	1	1	1	0	1	1	0	1	1	1

An eminently practical example: a switch debouncer

Refer back to page 23 where the problem of switch bounce was introduced. Simple logic gates can be used to make a switch debouncer. One solution (but not the only possible one) is shown in Figure 4.18.

If the switch is thrown down, input D of the lower of the two NAND gates is at logic low and holds output Q_2 at logic high. Both inputs to the upper gate are then high and output Q_1 is at logic low. When the switch is first opened, input D of the lower gate bounces but outputs Q_1 and Q_2 do not change since the low state of Q_1 holds input C of the lower gate at logic low and hence holds

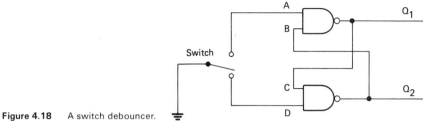

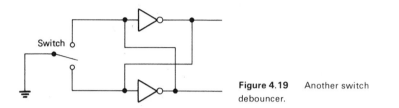

Figure 4.18   A switch debouncer.

Figure 4.19   Another switch debouncer.

output Q_2 at logic high. The outputs ignore the switch bounce when the switch is opened. When the switch begins to make contact with the lead to input A of the upper gate, the switch bounce is again ignored, while output Q_1 goes to logic high and output Q_2 goes to logic low. You are left to carry out the reasoning to determine that this is so.

Either of the outputs Q_1 and Q_2 could be used to drive an input pin on some TTL IC device. As the switch is thrown up and down, either output alternates between its low and high states. However, Q_1 and Q_2 are always in mutually complementary states. Another switch debouncer which is somewhat simpler than the one in Figure 4.18 is shown in Figure 4.19.

Three-State Logic

Suppose that in some digital electronic circuit it becomes necessary to connect the output pins from two different gates to one input pin of a third gate. This is illustrated in Figure 4.20 where outputs Q_1 and Q_2 of gates 1 and 2 are connected to input A of gate 3.

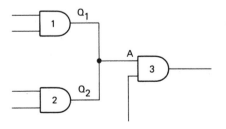

Figure 4.20   A circuit that depicts the need for a three-state gate.

If outputs Q_1 and Q_2 from gates 1 and 2 were coordinated so that both were high *or* low simultaneously but that it never happened that one is high and one is low, then all would be well. But if Q_1 and Q_2 operate independently, there is obvious trouble whenever they try to achieve opposite logic levels. $Q_1 = 0$ and $Q_2 = 1$, or $Q_1 = 1$ and $Q_2 = 0$ will not provide the input A to gate 3 with unambiguous information and, indeed, damage to the gates might result.

This kind of problem arises in complex systems such as computers where binary 0's and 1's are transmitted from various sources to some destination over a single *bus* line. For example, if output pins from 10 different ICs are to be connected to a single bus which leads to one input pin of some other IC, the situation shown in Figure 4.20 is multiplied greatly in seriousness.

A solution is provided by the *three-state gate*. This is a buffer for which the output state is not only a 0 or 1 but may in addition be *open*. When the output state is open, the output is decoupled from any circuitry to which it may be connected. A special input pin called *enable* (or sometimes *command* or *inhibit*) is used to determine whether the output pin of the unit is or is not coupled to the following circuitry.

Discussion of two particular instances should make this clear. Figure 4.21(a) shows the conventional symbol for a noninverting three-state buffer. How it works is shown in the table of part (b). When enable is high, the unit passes the input logic level on to its output, and when enable is low the output is open.

Figure 4.22(a) shows an inverting three-state buffer. How it works is shown in the table of part (b). This device is "low enable" while that which obeys Figure 4.21(b) is "high enable." (Notice the small circle where the enable line goes into the triangle in Figure 4.22.)

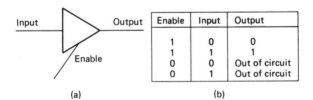

Enable	Input	Output
1	0	0
1	1	1
0	0	Out of circuit
0	1	Out of circuit

(a)   (b)

Figure 4.21   (a) Symbol and (b) truth table for a noninverting three-state buffer, high enable.

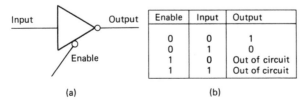

Enable	Input	Output
0	0	1
0	1	0
1	0	Out of circuit
1	1	Out of circuit

(a)   (b)

Figure 4.22   (a) Symbol and (b) truth table for an inverting three-state buffer, low-enable.

One popular TTL IC package of three-state units is the 74125 and another is the 74126. The 74125 is low-enable and the 74126 is high-enable. The pin diagrams are shown in Figure 4.23.

When a three-state gate becomes out-of-circuit on a suitable command at its enable input, the gate output is not *truly* disconnected from the line to which it is connected. Rather it is in a very high though not infinite impedance (resistance) state. In practice the effect is virtually the same as being actually out of the circuit.

Let us return to the problem presented in Figure 4.20. The way in which two units of a quad three-state buffer can be used to solve the problem is shown in Figure 4.24. If suitable signals to the enable pins of the three-state gates are provided, driven gate 3 receives an input signal at A only from output Q_1 of gate 1 or only from output Q_2 of gate 2 at any one time. To do this requires that the two enable signals be complementary. Hence they are labeled ENABLE and $\overline{\text{ENABLE}}$.

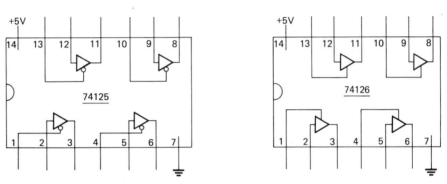

Figure 4.23   Pin diagram of the 74125 low enable noninverting three-state buffer and of the 74126 high enable noninverting three-state buffer.

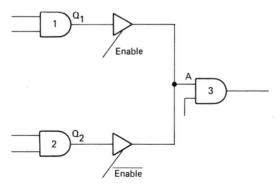

Table 4.24   Use of three-state devices to solve the problem in the circuit of Figure 4.20.

1. The two-line to four-line decoder

Figure 4.17 shows the nature of a two-line to four-line decoder. If you wish to build this circuit, you can use a 7404 hex inverter and a 7402 quad NOR gate to furnish the gates needed.

A more ambitious experimenter can undertake to devise a circuit *other* than that of Figure 4.17 which will perform the *same* functions, and build it.

2. Implementing an arbitrary truth table

Refer to Table 4.4, which presents a certain truth table, and to Figure 4.18, which shows a circuit that behaves according to that truth table. We suggest that you design from scratch a circuit which will behave according to some other two-input, one-output truth table. Either of the third and fourth truth tables of the four given in Table 4.5 can be used as the basis for this experiment.

3. Experiment with three-state buffers

Among the benefits that can be derived from this experiment are gaining familiarity with three-state devices, and with the basic concept of *multiplexing*. Multiplexing is an idea discussed fully in Chapter 11, and while you may choose to look ahead to that chapter to learn more about multiplexing now, it is not necessary that you do so to understand this experiment completely.

Refer to Figure 4.25. There are three inputs to the circuit: A, B, and C. A and B go both to the OR gate and to the NAND gate. Since the logic functions of these gates are different, the gates "process" the input information differently. For some A and B logic states, gate outputs Q_1 and Q_2 may well be different. If, for example, $Q_1 = 1$ and $Q_2 = 0$, then connection of Q_1 and Q_2 directly to the detector, rather than through the three-state buffers, would result in a conflict. Not only would the detector be confused concerning what logic state to show, but actual damage to some circuit component could occur.

Figure 4.25   Circuit for use with Experiment 3.

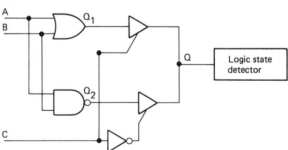

The two three-state buffers are in the circuit to prevent this by allowing one and only one of Q_1 and Q_2 to drive the detector. Furthermore, which of Q_1 and Q_2 is to be effective is to be determined at will by the experimenter by setting the logic state of input C. If, for example, the command signal at input C couples Q_1 directly to the detector, then Q_2 is to be automatically decoupled from the detector.

The purpose of the inverter is to make it possible to guarantee that when the state of input C is set, one and only one of the buffers will be "in" the circuit, and that the other buffer will be "out."

When the circuit has been built, changing the states of A and B while the state of C is held fixed will make it possible to determine by watching the detector which of the gates (OR or NAND) is connected to the detector, and so to check on the correctness of the circuit.

This circuit requires four IC packages. The OR gate can be one of those in a 7432, the NAND one of those in a 7404, the inverter one of those in a 7404, and the three-state buffers two of those in a 74126. The gates used need not be an OR and a NAND, but they should be different gates. Also, inverting buffers rather than noninverting buffers can be used.

The detector can be a logic proble (see the Logic State Indicators section of Appendix C), a LED with a suitable current-limiting resistor, or a suitable DC voltmeter.

Practical
TTL Considerations

Up to this point our emphasis has been on the underlying logic of the basic gates. In this chapter we will turn our attention to technical details about TTL ICs. Some of these are of interest but not ordinarily crucial. An example is the range of temperature over which TTLs can operate. Others are so important that you *must* know about them. An example is the difference between current sourcing and current sinking. In any event, you ought to be aware of what kinds of technical specifications *may* at some time be important to you as a TTL user, and you should also know the terminology, so that terms such as "fanout" or "Schottky" do not baffle you.

Power Supply Voltage

In our earlier discussions and circuit diagrams we have used the value 5 volts for the voltage required of the DC voltage source which powers TTL ICs. Actually there is some tolerance of variation from the value 5 volts. Voltages in the range from 4.75 to 5.25 volts are acceptable.

It is possible to power a circuit which uses TTL ICs with batteries if this is done properly. Four 1.5-volt dry cells and a silicon diode in series will provide a voltage fairly close to 5.25 volts. The forward voltage drop in the diode will reduce the nominal 6 volts of the cells by approximately enough. (In our experience, to use four cells without the diode is almost to guarantee ruining a

TTL device.[1]) This simple kind of power supply is usable only with small experimental layouts. The substitution of a 6-volt lantern battery for the dry cells would give some improvement. A 5-volt zener (regulator) diode used as a voltage regulator is also a feasable idea although the amount of circuitry such a supply can drive is very limited. The two kinds of simple power supply we have discussed here are illustrated in Figure 5.1.

The only completely satisfactory power supply is an electronic rectifier driven from the 110-volt AC lines, equipped with heavy filtering to remove ripple, and with a 5-volt regulator at the output of the unit. It must be able to supply all the current needed. Such power supplies are discussed in the first section of Appendix C.

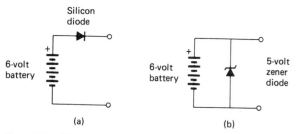

Figure 5.1   Two simple power supplies for TTL IC experimentation.

Logic Level Voltages

Earlier in this book we said that the voltage at a TTL input or output pin is 5 volts if the pin is at logic 1 and 0 volts if the pin is at logic 0. There could hardly be any TTL electronics if those statements were strictly true since such stringent requirements could not be met in practice. Actually there is a great deal of tolerance of variations from those values.

An input pin of a TTL device will recognize as logic level 0 any voltage from 0 to 0.8 volts and will recognize as logic level 1 any voltage from 2.0 volts upward. A TTL output pin provides a voltage which may be as high as 0.4 volts when it is low and a voltage as low as 2.4 volts when it is high. You should notice that the bounds on the voltages from an output pin are more restricted than are the bounds on the voltages which are required for recognition as logic 0 or 1 by an input pin, as must be true for reliable operation.

---

[1]Manufacturers usually state that 7 volts is the absolute upper limit that TTL can stand without damage.

Noise Immunity

Suppose that an input pin were to be brought to logic low with (say) 0.4 volts DC but that there were spikes of electrical noise present in the input signal. If the noise was large enough to bring the input pin to more than 0.8 volts momentarily, the device could interpret the signal as a logic high signal. A similar problem could arise if a noisy signal tried to hold a pin high.

The smaller the voltage region between the top of the logic 0 level and the bottom of the logic 1 level, the more immune the device is to electrical noise. The width of this region is a measure of the noise immunity. For TTL it is about 1.2 volts. Generally speaking this is good noise immunity but not nearly as good as with CMOS devices, about which more will be said in Chapter 15.

Current Requirements

About 18 milliamperes of current are consumed by simple logic gates on the average, including both the standby current (which the internal transistors draw even while the device is not processing data) and operating currents. This is a typical figure regardless of the frequency of the signal. More complex TTLs may draw 50 milliamperes or more.

There are two more data about TTL currents which are of special interest. One is that the maximum current which an output pin can deliver to a load is about 18 milliamperes and the other is that input pins typically draw about 1.8 milliamperes.

Fanout

Not uncommonly one output pin of a TTL device must be used to drive input pins on several other TTL devices. This situation is represented schematically in Figure 5.2. The number of inputs which can be driven by one output pin is called the *fanout*.

Figure 5.2   One IC output driving several IC inputs.

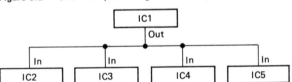

You would naturally like to know how many inputs can be driven by one output. Since the current from an output is 18 milliamperes and each input requires 1.8 milliamperes, maximum TTL fanout is 10.

Current Sourcing and Current Sinking

Refer to Figure 5.3. An output pin of a TTL device is connected to some external load which might be a resistor or an indicating lamp, for example. The other end of the load goes to the circuit ground. Let us suppose that the load is a buzzer made for TTL operation and that it draws about 2 milliamperes of current at about 5 volts. When the output pin is at logic level 0, the buzzer will not sound. When the output pin goes to logic level 1, there will be 5 volts or a bit less across the buzzer, and it will sound. What is important in this instance is that the IC which drives the buzzer is the *source* of the current drawn by the load. This is a case of *current sourcing.*

Another way to drive the load is shown in Figure 5.4. In this case, the TTL output pin must go *low* in order for the load to be activated. When the pin goes low, internal circuitry makes a connection from the load through the TTL circuitry to the circuit ground. (The internal path to ground is not shown in the figure.) This completes a current path from the positive terminal of the external power supply through the load to the circuit ground. This allows current to flow and to activate the load. This mode of operation is called *current sinking.* The TTL must *sink* the current rather than *supply* it. TTL devices are available which are able to sink substantial current. For example, a selected TTL chip may be able to sink the 200 milliamperes demanded by a small incandescent lamp while an output pin used in current sourcing mode could not supply that much current.

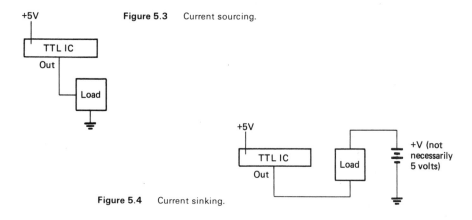

Figure 5.3   Current sourcing.

Figure 5.4   Current sinking.

Power

When a DC voltage V drives a current I through a load resistance R, the power expended is equal to the product of the current and the voltage: $P = IV$. When V is in volts and I is in amperes, P is in watts. If we take V to be 5 volts, our formula reduces to $P = 5I$ watts. If the average current was 25 milliamperes (an amount between the figures of 18 and 50 milliamperes noted earlier in this chapter), the average power expenditure would be 125 milliwatts (⅛ watt). With this estimate of the power per IC in some average sense, you can get a reasonable estimate of the total power requirement of a set of ICs. For example, a set of 10 ICs would use about 1.25 watts. Of course, the power supply should have plenty of "headroom" above this possibly conservative estimate.

Speed and Frequency

When the state of an input pin is changed from 0 to 1 or from 1 to 0, the corresponding change at an output pin does not occur instantaneously. This is so because electrons must be moved physically in the internal transistors to bring about the response and this takes some time. The time required is known as the *propagation time*.

For *regular* TTL gates, the propagation time is about 10^{-8} seconds, or 10 billionths of a second, but this is nevertheless a finite time which becomes important in high-speed operations. (The distinction between "regular" and other varieties of TTL will be discussed in the section of this chapter on Varieties of TTL.)

If you assume that the maximum frequency of operation is the reciprocal of the propagation time, as seems plausible, then the maximum frequency for gates would be about 100 megahertz. While gates may indeed approach this value, they are most often used in conjunction with other devices such as flip-flops. Flip-flops are the main elements in counters, arithmetic units, and memories. The maximum frequency for a counter in regular TTL form is only about 35 megahertz, and this sets the upper limit on the system operating frequency.

A rule which often works out well in many cases is that the maximum operating frequency is about one third of that which the reciprocal of the propagation time would suggest.

Temperature Limitations

TTL devices in the 7400 series should be operated in the temperature range from 0 to 70°C (32 to 158°F). While the upper temperature is fairly high [70°C (158° F)], the lower temperature limit [0°C (32° F)] is quite moderate. When the device temperature approaches the limits of its range, properties such as maximum speed frequency of operation can deteriorate considerably.

TTL devices in the 5400 series tolerate a much wider range of temperatures, from −55 to 125°C (−67 to 257°F).

Varieties of TTL

Regular TTL

The 7400-series TTL which we have discussed thus far were the first TTLs produced by the industry. Let us call them *regular* TTLs to distinguish them from other varieties of TTL which we are about to describe. They are cheap, reliable, readily available, and have other desirable properties, as we have seen. Also, devices performing a wide range of functions are made in 7400-series form. Here is some principle data for regular 7400-series TTL.

Type[2]	Gate Propagation Time[3] (seconds)	Power per Gate (milliwatts)	Maximum Flip-Flop Frequency (megahertz)
74XX	10^{-8}	10	35

[2]74XX is a brief way to suggest such numbers as 7400, 7404, 74193 and others.

[3]The times shown are for NAND gates. The times for devices which perform other functions may vary considerably from these.

Low-power TTL

Another kind of TTL is known as *low-power TTL*. This series bears numbers according to the format 74LXX. As the name suggests, low-power TTL offers the advantage of significantly less power consumption than for regular TTL.

Type	Gate Propagation Time (seconds)	Power per Gate (milliwatts)	Maximum Flip-Flop Frequency (megahertz)
74LXX	3.5×10^{-8}	1	3

74LXX TTLs are becoming extremely popular because of their low power consumption despite some shortcomings in them. One disadvantage (not shown in the display of data above) is that the fanout is only 2 compared with the fanout of 10 for regular TTL.

Another disadvantage is the low speed of operation. This is an instance of a general rule which can appropriately be brought up at this point: In the current state-of-the-art there is usually a tradeoff between power consumption and speed of operation. The operation of the rule is evident when you compare regular and low-power TTLs. This rule also operates with respect to TTL and CMOS devices, the latter consuming much less power than do any TTL devices, but at the cost of slower operation. This is discussed further in Chapter 15.

You should note that not all of the logical functions available in regular TTL form are to be had in low-power TTL form although many are.

High-power TTL

There is a variety of TTL which is capable of high speed but at the cost of higher power consumption than for regular TTLs. This is *high-power TTL.*

TTL Type	Gate Propagation Time (seconds)	Power per Gate (milliwatts)	Maximum Flip-Flop Frequency (megahertz)
74HXX	0.6×10^{-8}	22	50

Not all functions are available in 74HXX form.

Schottky TTL

Schottky TTLs are named after Walter Schottky, a designer who found a way to improve the speed performance of TTLs by incorporating certain special diodes (Schottky diodes) into the transistor circuitry. The improvement is evident in the display of data below.

TTL Type	Gate Propagation Time (seconds)	Power per Gate (milliwatts)	Maximum Flip-Flop Frequency (megahertz)
74SXX	0.3×10^{-8}	19	125

The power consumption of Schottky TTLs is rather high. Also, there are rather few regular TTL devices which are also available in Schottky form.

Low-power Schottky

Low-power Schottky devices combine some of the high speed of the Schottky type with the desirable property of low power consumption.

TTL Type	Gate Propagation Time (seconds)	Power per Gate (milliwatts)	Maximum Flip-Flop Frequency (megahertz)
74LSXX	0.95×10^{-8}	2	45

Not all functions are available in 74LSXX form.

Some General Comments

For most work regular TTLs are fully adequate. They are by and large the most readily available and usually they are the least expensive. Applications where they are satisfactory include general experimentation and a large proportion of circuits for practical use. In portable battery-operated digital devices, power consumption is obviously a very important consideration. High-speed computer and memory systems present special problems which influence the choice of kind of digital ICs. The circuit designer or builder must weigh the technical demands of what is contemplated against such practical considerations as cost. As for cost, we present Table 5.1 in which we show that we have learned from a survey of dealer's catalogs.

Table 5.1 Typical costs of some of the varieties of TTL devices.

TTL Type	Representative Cost per TTL Package
7400	$0.19
74L00	0.22
74H00	0.25
74S00	0.59
74LS00	0.35

Open Collector TTL

There is one more distinctive kind of TTL device to be discussed. It is called *open collector* TTL. Because this kind of TTL has some very special characteristics and is not uncommon, it is important that you have an understanding of it.

In this book we almost entirely avoid details of the interior transistor electronics in digital ICs. However, it is not possible to discuss open collector TTL without delving a bit into circuit details. Figure 5.5 shows a part of the circuitry associated with an output pin in the case of most TTL ICs and Figure 5.6 shows the corresponding fragment of circuitry associated with an open collector output pin.

In the "totem-pole" arrangement of Figure 5.5, the collector of the lower transistor T_2 goes to the output pin but it also goes through a diode (D) to the

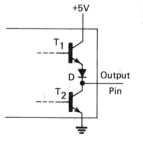

Figure 5.5   The totem-pole ouput configuration.

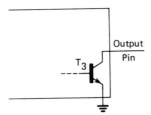

Figure 5.6 The open collector
output configuration.

emitter of the transistor T_1 above it. These two transistors have connections to still other transistors and other components elsewhere in the circuitry (omitted in the drawing). In contrast, the open collector device in Figure 5.6 has the collector of the output transistor T_3 connected only to the output pin, hence the name open collector.

In almost all of our discussions of ICs thus far we have had totem-pole TTLs in mind. (One exception we have met is the 7407.) We have concentrated on how to use this kind of TTL and on how it behaves. We turn now to look at some of the properties of open collector TTL devices.

To connect a load such as an LED or an input pin of another IC to an open collector output pin you must connect the load to a transistor collector which has no other connection. In order to turn the output transistor (which is an NPN transistor) on, the collector must get electrical contact with a suitable positive voltage. It can do this through the load to an external power supply. In short, an open collector device is a current-sinking device.

Open collector devices are usually *drivers,* meaning that they can sink substantial current and stand substantial voltage at their output pins. Open collector inverters and noninverting buffers are common examples. The 7407 can sink 30 milliamperes of current and withstand 30 volts at an output pin. It is indeed a driver.

It is usually as easy to use an open collector device (such as the 7407) as it is to use totem-pole types. However, surprises can be in store for you if you are not aware that the device you are about to use is an open collector device. Furthermore, since the technical specifications for open collector ICs are different than those for regular TTL, the only safe procedure is to check the specification sheets or data books.

In the literature which deals with open collector devices, there is usually discussion of the "wired NOR" circuit which is possible with open collector outputs. This is not a matter of any practical importance for beginning students of digital IC electronics but we will explain it here for the sake of its contribution to understanding open collector devices.

First consider Figure 5.7. The inverters shown are assumed to be ordinary TTL gates. There is conflict if one of the outputs tries to go high and one or more of the other outputs try to go low. The result might be destruction of some of the hardware. Then consider Figure 5.8 where the gates are of the open collector kind. If any one of the gate outputs goes low, then all of the output collectors will go low and the output transistors will be off. No conflict can arise. A system which behaves in the way we have just descibed is a NOR. We say that the arrangement of Figure 5.8 is a wired NOR.

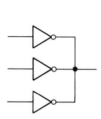

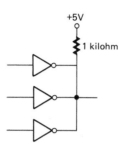

Figure 5.7   Outputs from three inverters wired together.

Figure 5.8   The wired NOR: Outputs from three open collector inverters wired together.

Oscillators

In the world of electronics in general, oscillators have an importance that cannot be exaggerated. To mention their use in radio and television transmission and reception would be hardly more than to hint at their many applications. In digital electronics, oscillators are most often called *clocks* and examples abound in which clock signals are essential. These include calculators and computers, digital watches, frequency counters, music synthesizers, and control devices. This chapter will explore the use and design of simple IC clocks. Although this is a book about digital integrated circuit devices, we include a discussion of the enormously popular 555 timer.

A Simple Introductory Example

Prior to the advent of electronic timing the world's fastest runners were timed by hand-operated stopwatches, usually using several stopwatches and taking an average. This method is hardly satisfactory for the precise timing of world records. The digital stopwatch schematically displayed in Figure 6.1 can be exceedingly accurate while still being very simple.

The clock produces a regular sequence of pulses at some rate such as (say) 10,000 per second. The "counter" accepts these pulses (once turned on) and counts them. At any instant the current count is a measure of the elapsed

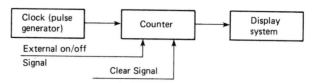

Figure 6.1   Schematic of some of the essentials of a digital watch.

time since turn-on. When upon command the counter stops counting, the total count is a measure of the total elapsed time. The "display system" manages the job of displaying the final count (total time) in nicely readable form as decimal numbers complete with a decimal point. The "external on/off signal" can be obtained automatically by some devices at the starting and finishing points of the run. The purpose of "clear" is to reset the display panel reading to zero when a new run is to be timed.

The nature of the *counter* and the *display system* will be discussed in later chapters. Our concern now will be with a variety of ways in which the clock can be made.

Simple Gate Clocks

First we offer an analysis of the proposed, but not actually practical circuit of Figure 6.2. Our purposes are to convey some basic ideas and to teach a moral.

The key to understanding the dynamics of the circuit is in the charging and discharging of the capacitor along with the consequent changes of the state of the output. Let us assume that the capacitor C is initially charged as shown in Figure 6.3. Then all the input and output states of the NOR gates are determined. These states are shown in the figure. The capacitor will discharge through circuitry in the interior of NOR gate 1. At some time during the discharge the output of NOR gate 1 will no longer be at logic high and the uppermost of the inputs to NOR gate 2 will no longer be at logic high. Then

Figure 6.2   A pulse generator or clock.

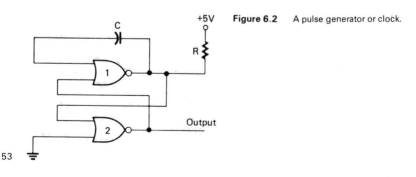

53

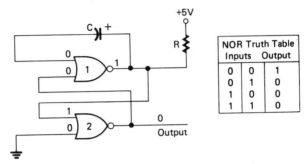

Figure 6.3 Some of the dynamics of the clock of Figure 6.2.

NOR Truth Table		
Inputs		Output
0	0	1
0	1	0
1	0	0
1	1	0

NOR gate 2 changes its output from 0 to 1. (The NOR truth table shown in the figure can help in seeing that.) The lower input of gate 1 changes from 0 to 1. To correspond to the truth table, the upper input of gate 1 goes to state 1.

The transitions described are indicated in Figure 6.4. There has been a reversal of the two gate outputs and, in particular, of the state of the output of gate 2 which is the oscillator output. The capacitor then charges up again and eventually the charge polarity will be that shown in Figure 6.3. The output will go to 0 once more. The cycle will recur over and over.

Actually, this circuit will not work with TTL gates because the internal TTL electronics are not compatible with the proposed functioning. Moral: Life on paper and life in the real world may be different.

In Figure 6.5 is shown a related, and eminently practical, TTL clock. It is left to the reader to analyze how this clock works.

Figure 6.4 Dynamics of the clock of Figure 6.2, continued.

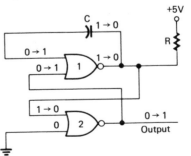

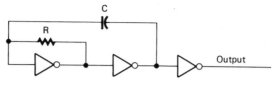

Figure 6.5 A TTL clock circuit.

The 555 Timer

In recent years a device known as a *555 timer* has become very popular for making a clock for use with TTL or other circuits. The 555 has many other applications as well and the interested reader may wish to consult *TTL Cookbook*[1] or the more elaborate *IC Timer Cookbook.*[2]

With only a few electrical components added, the 555 furnishes a clock which is adequate for most applications. It can be powered with the usual 5-volt DC supply and its output is then compatible with TTL input needs. It is readily available and is inexpensive. The clock frequency can be adjusted to range from virtually zero up to about 100,000 hertz.

The 555 is available in an 8-pin DIP format with pin spacings similar to those of 7400-series TTL devices so that it fits standard breadboards and sockets. It is also available in a dual form (two 555s), which is then known as a 556. (The 555 is often offered by dealers in the form of a cylindrical can with eight wire leads. This package is inconvenient to use when the rest of a circuit is in DIP form.)

The form of the 555 DIP and the names used to indicate the functions associated with the pins are shown in Figure 6.6.

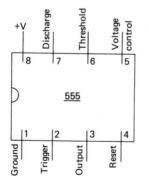

Figure 6.6 The pin functions of the 555 timer.

---

[1] *TTL Cookbook*, by Don Lancaster, Howard W. Sams and Co. Indianapolis, IN, 1974.

[2] *IC Timer Cookbook*, Walter G. Jung, Howard W. Sams and Co., Indianapolis, IN, 1979.

Figure 6.7 indicates how two resistors, a capacitor, and a power supply can be added to the 555 to make a clock for TTL applications. We assume that when the circuit is first turned on the output pin (3) is at logic high and that the capacitor C is uncharged. The capacitor proceeds to charge up through resistors R_1 and R_2. In time, this action brings pin 6 (threshold) to 3 volts. This causes the internal circuitry to switch the output pin (3) to logic low. Simultaneously, circuitry associated with pin 7 (discharge) makes it possible for the capacitor to discharge through resistor R_2. As the capacitor discharges, the voltage at pin 2 (trigger) drops. When this voltage becomes 1⅔ volts, internal action sends the output pin (3) back to its logic high state. Thereupon the capacitor begins to charge through R_1 and R_2 and the cycle repeats continuously.

Two important considerations which have to do with this nice circuit are what frequencies and waveforms are obtainable. Suppose that you as an experimenter want the 555 clock to drive an experimental circuit at such a low rate that you will be able to follow the circuit action. For example, you may want to watch the seven-segment displays driven by a counter which, in turn, is driven by the clock. With R_1 and R_2 equal to about 6000 and 1000 ohms respectively and with C equal to about 500 microfarads, the output signal at pin 2 will have a frequency of about ½ hertz. With R_1 and R_2 unchanged, but with C equal to about 100 microfarads, the frequency will be about 3 hertz. With either of these values for the external components in the circuit, the waveform will be like that shown in Figure 6.8.

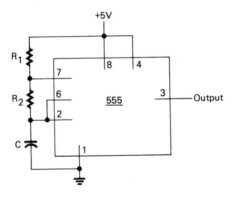

+5V

R₁

R₂

C

8   4

7

6   555   3 — Output

2

1

Figure 6.7   The 555 used to make a clock.

Pin 3
voltage

Time

Figure 6.8   Typical output waveform of the 555 timer wired as a clock.

In such a case, the output is at logic high most of the time, dipping periodically to logic low. The *duty cycle* (ratio of time high to the cycle time) is somewhat close to 1. If the waveform was perfectly symmetrical, the duty cycle would be ½. A symmetrical output waveform can be created by suitable choices of the circuit components.

The frequency of oscillation can be adjusted over a wide range by changing the values of the components. The frequency is given by the expression $(1.46/(R_1 + 2R_2)(C).$[3] Further details and graphical aids in choosing component values are given in the references cited earlier in this chapter where discussion may also be found about the waveforms.

The 555 timer has excellent characteristics. It is able to handle load current demands at the output pin (3) up to about 200 milliamperes when the power supply voltage is 5 volts. Under no-load conditions the 555 draws only about 3 milliamperes (at 5 volts) or only about 15 milliwatts of power. It can be used from about 0 to 70°C (32 to 150°F) and, while the frequency of the oscillator circuit may drift when the temperature changes, the drift is only about 150 parts per million per degree Celsius. Finally, as we have already said, the output is compatible with the needs of the inputs of TTL devices if the 555 is powered with a 5-volt supply.

EXPERIMENTS

1. A clock based on a 555 timer

The purpose is to construct a clock, using a 555 timer. This circuit is a permanently useful one. It should be built with reasonable compactness, leaving room on the breadboard for later circuit building.

Install a 555 timer on the board. Then build the circuit shown in Figure 6.9. The numerical values of the resistances and of the capacitance are chosen to give a clock rate which is somewhat less than 1 hertz so that when your timer is used to drive other circuit elements such as counters, the action can be followed with your eyes. The values of the components can be varied substantially without significant effect on the clock rate, so that any components you have on hand which are reasonably close will work. (If you have a 556 timer, which is a dual form of the 555, the proper pin numbers must be determined before building the 555 circuit.)

---

[3]The frequency will be in hertz if each resistance is in ohms and C is in farads. This formula gives only roughly correct predictions in the region of a few hertz, but serves better in the region of several kilohertz to a few megahertz.

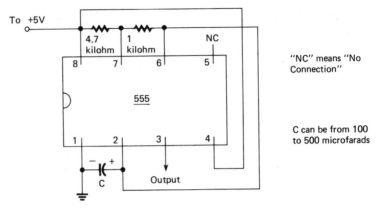

To +5V

4.7 kilohm   1 kilohm   NC

8   7   6   5

555

1   2   3   4

C

Output

"NC" means "No Connection"

C can be from 100 to 500 microfarads

Figure 6.9   Circuit details of the low-frequency clock of Experiment 1.

The capacitor used probably has polarities marked on it. It may be important to observe these polarity markings according to Figure 6.9, since the circuit may not oscillate otherwise.

In order to verify that the clock is working, connect a logic state indicator between pin 3 of the 555 and the circuit ground. The indicator can be a suitable DC voltmeter or a discrete LED. An oscilloscope can be used.

It should be apparent from the response of the indicator that this clock is high most of the time, dropping periodically for short time intervals to low. This form of clock output will serve well for the experiments which will be described in later chapters of this book.

A simple experiment which is well worthwhile is changing the frequency of the clock. This is easily done by changing the value of capacitor C. The waveform of the output signal can also be experimented with although this requires an oscilloscope.

2. A simple TTL clock

Figure 6.5 shows a simple clock which uses TTL inverters and a few other components. It is easily built and easily studied with simple output state detectors if the clock is made to operate at very low frequencies. At frequencies too high for the human eye to follow, an oscilloscope is essential for experimenting with these. Such matters as the details of the output waveform and the duty cycle also require use of an oscilloscope. Nevertheless, you may want to build this clock since little time is required to do so, in order to get some feel for how it works and in order to contrast it with the 555 timer clock in Experiment 1.

3. A more elaborate TTL clock

We speak now to the reader who has had more experience in building electronic circuits than we have previously assumed in this book, and who intends to go further into the world of IC electronics. Such a reader may want to consider building a more sophisticated clock for general experimentation, probably mounting it up into a chassis to make it a laboratory instrument.

Figure 6.10 shows a clock we have used. It features variable output frequency and variable duty cycle. With the low/high switch set at low the frequency can be set from about 1 hertz to about 10 hertz by adjusting the frequency (10-kilohm) potentiometer. With the switch set to high the frequency range is from about 1 to 10 kilohertz. The other 10-kilohm potentiometer adjusts the duty cycle. The output signal has quite sharp rising and falling edges. The gates are all contained in a single 7400 quad NAND package. Pins 7 and 14 (not shown) go to ground and to +5 volts, respectively. The diodes improve the waveform. Switching diodes should be used. The 33-nanofarad capacitor shunts transient switching spikes to ground.

Such clocks and others with wider frequency ranges and other capabilities are available on the market, but of course their prices are correspondingly high.

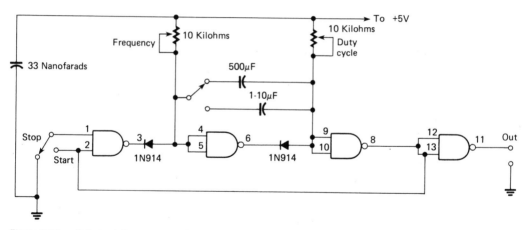

Figure 6.10   Full circuit for a more sophisticated clock using TTL devices.

7

Monostables
and Schmitt Triggers

Multivibrators: Terminology and Some Applications

Each of the clock circuits described in Chapter 6 has the property that its output oscillates between its high and low states. It is not stable in either output state which is precisely why it is useful. Any such circuit is called an *astable multivibrator* or sometimes a *free-running multivibrator*. The word "multivibrator" is inherited from the realm of vacuum tube and transistor electronics. It refers to a family of three distinct kinds of circuits which includes astables, bistables, and monostables.

A *bistable multivibrator*, much more commonly called a *flip-flop*, is a circuit which is stable with its output high or with its output low, but which can be switched from either output state to the other by command by an externally applied *trigger* signal. After the trigger signal has caused the flip-flop to establish its output state, that state is maintained steadily until eventually another trigger signal causes it to switch to the other output state. The name "bistable" is appropriate, as is the name "flip-flop." Flip-flops are of such importance in digital electronics that the matter cannot be exaggerated. We devote all of Chapter 10 to them.

A *monostable multivibrator*, or simply *monostable*, is also often called a *one-shot*. A monostable normally holds its output high or low, depending on the type. This is its one ("mono-") stable state. However, on command of an externally applied trigger signal, the monostable will switch to its other output state, hold its output state there for a certain time and then will revert to its one stable state. Let us suppose for the sake of definiteness that a particular mono-

60

stable has its output normally low. When triggered the device output will switch to high, stay high for a while, revert to low and wait at low for the arrival of another trigger signal.

A monostable can be useful when the generation of a voltage pulse of preset duration is needed. Frequently in applications monostables are used to delay some circuit action. Consider an experiment in which a large amount of noise is always present at the outset. You could delay the data-taking by an appropriate time by use of a monostable. In a cathode-ray oscilloscope when the screen display sweep is triggered by a signal, the details of the early part of the display may be hard to see on the screen. If the sweep is started at once by the sweep-trigger signal but the vertical component of the display is delayed for a short time after the sweep begins, all may be much clearer. This also is a situation which suggests the use of a one-shot as a delay device.

A time delay can be introduced into the action of a circuit by the use of certain other devices called *shift registers* and *counters*. In many instances it is preferable to use those devices, since monostables are essentially dependent on external capacitors and resistors for their action, while shift registers and counters are not and can be more precise and more reliable. Nevertheless, some familiarity with monostable multivibrators should be part of the background of all students of digital electronics.

Monostables Made With Gates

A one-shot can be made with a remarkably simple combination of logic gates. One way is shown in Figure 7.1.

Suppose that input A of the circuit is normally low, due to some other circuitry which is not shown. You see that the output Q is therefore normally high. Suppose that a signal on the lead to A brings A high, holds A high momentarily, and then drops A back to low. This signal pulse has a *rising edge,* a *level* of some duration, and a *falling* edge. In response, output Q of the circuit, which is normally high, will fall to low, remain at low for some time and then revert to high (in the manner of one-shots). This output.pulse will have its rising edge at a time later than the time at which the rising edge of the trigger pulse arrived at A. This constitutes the delay action of the one-shot.

In order to expedite our next discussion, that of the action of Figure 7.1 circuit, Figure 7.2 is presented.

Figure 7.1   A one-shot made with an inverter and a NAND.

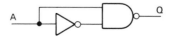

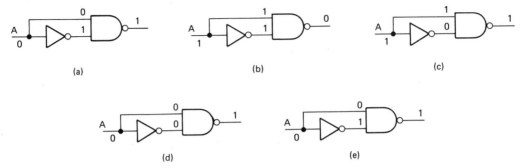

(a) (b) (c)

(d) (e)

Figure 7.2 Analysis of the action of the circuit shown in Figure 7.1.

In this discussion we will assume that there is a substantial propagation delay in the inverter and little delay of any significance in the NAND gate. This assumption is made to help provide a clear description of the essence of how the circuit works. An interested reader may wish to contemplate the effect of the propagation delay in the NAND gate on the circuit action.

In Figure 7.2(a) all of the states are in their normal conditions. In part (b) the rising edge of an input pulse has occurred. The upper input to the NAND instantly has changed to high, but the lower input remains high (as it was) because the time required for the inverter to switch its output has not yet passed. Output Q has switched from 1 to 0 in response to the rising edge of the input signal. In part (c) the inverter has just switched its output from 1 to 0, and in response the NAND has returned its output to 1. The rising edge at the output has appeared but at some time after the rising edge of the input pulse. This is the delay action of the circuit. In order to return the circuit to its original states as in part (a), the input signal must fall back to low. In part (d) this has just happened, and the inverter has not yet returned its output to 1. In part (e) some time has passed and the inverter output is now 1. Notice that in parts (d) and (e) output Q stays high, as it was in part (c) and as it is normally, so that the falling edge of the input pulse had no effect.

The time delay for this circuit cannot be expected to be long, since the inverter does not have a long propagation time. A simple way to stretch the time is shown in Figure 7.3. This makes use of the *time constant* of the resistor-capacitor combination. (The time constant is the product RC and is in seconds if R is in ohms and C is in farads.) The behavior of this circuit essentially is the same as that of the previous one, but the time required for the output of the

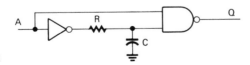

Figure 7.3 Resistor and capacitor added to the one-shot of Figure 7-1.

inverter to affect the state of the lower input to the NAND can be drastically affected by the RC combination.

TTL Package Monostables

A representative TTL monostable is the 74121. The pin diagram is shown in Figure 7.4. Note that an external capacitor and resistor must be used with the IC so leakage, imprecise internal values, and other vagaries may cause difficulties in critical applications. However, in general the device is simple to use and reliable.

The 74121 is *nonretriggerable*. This means that while the output is active the device will not respond to another trigger input pulse. Such a one-shot is sometimes said to be a *dead-time device*.

Pins 1 and 6 furnish conplementary outputs. There are three pins (3, 4, and 5) used for triggering rather than just one. These can be used in three different ways. One way is to have A_1 and A_2 low and to change B from low to high to produce triggering. Another way is to have A_1 and B high, and to change A_2 from high to low. Finally, you can have A_2 and B high, and change A_1 from high to low.

A companion to the 74121 is the 74122. This device is *retriggerable*. If a trigger pulse has activated the output, a second trigger pulse can be applied

Figure 7.4   Pin diagram of the 74121 monostable.

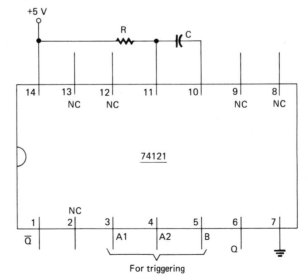

with the result that the duration of the output signal will be extended. Details of the 74122 are to be found in the manufacturer's specification manuals and elsewhere, as in the *TTL Cookbook* by Don Lancaster.

The 555 Timer Used as a Monostable

In Chapter 6 we pointed out that the 555 timer is very popular and is frequently used as a clock to drive TTL circuits. The 555 can be used to make a one-shot circuit. One way is shown in Figure 7.5.

The width of the output pulse (delay time) will be a function of R and C. A reader interested in details should have no difficulty in finding copious information in the literature.

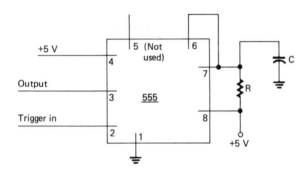

Figure 7.5   The 555 timer used in a one-shot configuration.

The Schmitt Trigger

The Schmitt trigger is a circuit which is related to one-shots in that it does not oscillate spontaneously but must be triggered into action. However, once its output is switched by an input signal, it does not recover spontaneously. It must be returned to its original state by a second signal. The two input signals are the rising and falling parts of some continuous pulse. This will be made clear in what follows.

The Schmitt trigger is used as a *pulse-conditioner*. It can convert a sloppy signal (one of irregular form) into a neat square wave shape which will be accepted with reliability as an input by any TTL IC. Consider as an example a digital electronic clock. The basic 60-hertz timing signal is assumed to be obtained from the AC line to which the clock is connected. This 60-hertz signal will be sinusoidal in form, as shown in Figure 7.6. It must be used as input to some TTL IC such as a counter. The TTL IC needs a relatively fast, snap-action signal at its input. A Schmitt trigger is placed between the source

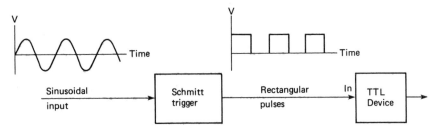

Figure 7.6   The Schmitt trigger used as a pulse-conditioner.

of the sinusoidal 60-hertz signal and carries out the pulse conditioning which is needed.

We start our discussion of the Schmitt trigger by representing it in the conventional form shown in Figure 7.7. As the symbol suggests, the Schmitt trigger is an inverter—high in, low out and low in, high out. The origin of the S-like symbol shown inside the inverter in the figure will be explained presently.

In Figure 7.8 the voltage at the input of the Schmitt trigger is plotted horizontally, and is assumed to start at zero, rise to +5 volts and then fall again to zero. Correspondingly the voltage at the output is plotted vertically and begins at +5 volts. When the initially zero input voltage rises to a trip point which is taken to be 0.9 volts, the output signal falls very abruptly from +5 volts to zero. As the input signal continues to rise to some maximum, which is taken to be +5 volts in the figure, the output remains at zero. When the input voltage thereafter falls to another trip point, which is taken to be 1.7 volts, the output abruptly rises to its original high level, where it remains as the input voltage continues to fall to zero.

Notice that the numerical values for the two trip points are not the same. This characteristic action of the Schmitt trigger is sometimes referred to as *hysteresis* because of its resemblance to the hysteresis behavior shown by an

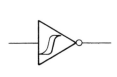

Figure 7.7   Symbol for the Schmitt trigger.

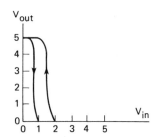

Figure 7.8   Output voltage versus input voltage for a Schmitt trigger.

iron sample as it is repeatedly magnetized and demagnetized. The curve in Figure 7.8 has the shape of an open S-like figure, which happens to appear right-to-left reversed. (If the Schmitt trigger in Figure 7.7 were not an inverter, the curves in Figures 7.7 and 7.8 would be directly comparable.)

In our discussion, the trip points were taken to be 0.9 and 1.7 volts. These values are correct for the 7414 Schmitt trigger. This frequently used device is shown in pin diagram form in Figure 7.9.

A significant use for a Schmitt trigger is in helping to clear up a noisy signal. Suppose that the input signal to a Schmitt contains some extraneous noise component. The output level of the Schmitt is for most of the time locked high or low, and the noise is ignored most of the time. The manner in which this works is suggested schematically in Figure 7.10.

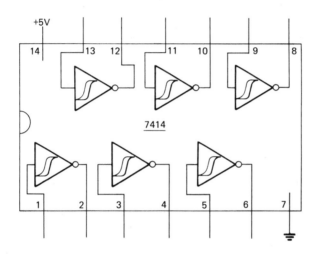

Figure 7.9   Pin diagram of the 7414 hex Schmitt trigger.

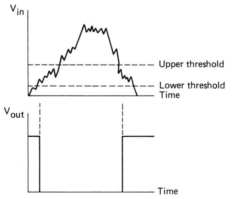

Figure 7.10   Schmitt trigger action in cleaning up a noisy signal.

It is possible to make a Schmitt trigger out of simple logic gates. The circuit in Figure 7.11 requires only a single noninverting buffer and two resistors. We have found that this circuit works with R_{in} equal to 480 ohms and R equal to 6.8 kilohms. You may wish to try other values, and if you have suitable apparatus, see if you can determine the thresholds.

In Figure 7.12 we show a way to make a Schmitt trigger which uses NAND gates and no external circuit components. The hysteresis of this circuit is determined by the inherent characteristics of the NANDs used. This circuit will be faster than the previous one although the thresholds are not adjustable. Since speed in the snap-action is what is usually wanted most, this NAND-gate Schmitt trigger may be desirable.

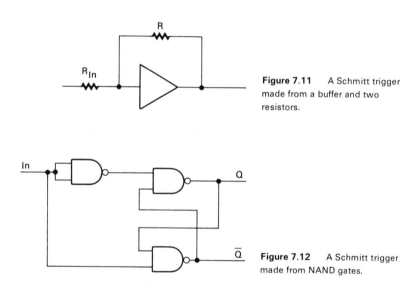

Figure 7.11   A Schmitt trigger made from a buffer and two resistors.

Figure 7.12   A Schmitt trigger made from NAND gates.

EXPERIMENTS

1. Monostables made with gates

It is desirable to make one or two monostables—and it is easy—in order to get a feel for their live behavior. A monostable with very slow action, which might be of the order of a few seconds, can be watched at its output with a simple LED indicator. If you have a cathode ray oscilloscope you can study much faster circuits and do more quantitative studies.

Figure 7.13 shows two suitable monostable circuits. These use RC combinations in different ways to affect the timing of the changes in the output state.

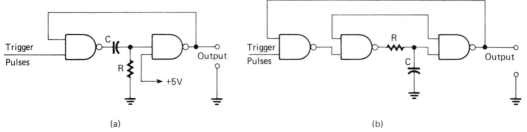

(a)   (b)

Figure 7.13   Monostables made with gates for use in Experiment 1.

The trigger (T) pulses should be from a debounced switch if manual pulsing is used. For long delay times, R can be 200,000 ohms or thereabouts and C can vary from about 0.01 to 5 or more microfarads. Notice that another monostable circuit is given in Figure 7.3.

2. A TTL one-shot
As we pointed out in the text, monostables in TTL packages are available. Figure 7.4 shows how a 74121 can be used. As in Experiment 1, the values of R and C can be varied. A study of the variation of the time delay as it depends on R and C might be considered by those with special interest in the matter. If the values of R and C specified in Experiment 1 are used, the action will be slow enough to be observed with a simple indicator at the output.

3. A 555 timer monostable
The popular 555 timer can be wired up to act as a monostable, as shown in Figure 7.5. This circuit has many applications, and is interesting to build and study. When the trigger pulse arrives, the capacitor C charges up to about ⅔ of the 5 volts supplied to the device, and then the capacitor discharges rapidly to ground potential, and the output at pin 3 remains high for a time T. The time T can be calculated using the relation $T = 1.1\ RC$, with T in seconds if R is in ohms and C is in farads. This can be checked experimentally once the circuit is built.

4. A Schmitt trigger experiment
The purpose in this experiment is to use a Schmitt trigger made from gates and a very few external components to convert a sinusoidal voltage input signal into an approximately rectangular output signal. An experimental arrangement is shown in Figure 7.14. The AC voltage source shown is a 6.3-volt filament transformer, which can be purchased off-the-shelf at Radio Shack outlets or from other dealers. If you have on hand a laboratory-type audio frequency signal generator, this can be substituted for the transformer with the advantage that both the frequency and peak voltage of the signal applied to the Schmitt trigger can be varied. Suitable values for R_{in} and R are about 480 ohms and 6.8

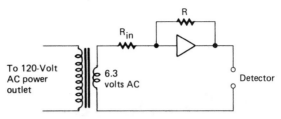

Figure 7.14   Circuit for observing action of a Schmitt trigger in converting a sinusoidal waveform into a rectangular waveform.

kilohms, respectively. At the power line frequency of 60 hertz, the switching of states at the output is too fast to be followed with the eye if a simple LED logic state indicator is used. An input signal with a frequency of 1 hertz or so must be used instead. Alternatively, you can use an oscilloscope to follow the action at the output, and to compare it with the input signal.

5. Experience with a TTL Schmitt trigger
A variation of the experiment above is to use a TTL Schmitt trigger such as the 7414 in a similar experiment. The TTL device replaces the gate, R_{in}, and R in Figure 7.14.

Counters, Decoders, and Displays

An important function performed by ICs in many devices is counting. In the device we examined at the beginning of Chapter 2 a counter was an essential part of the system described. Other applications occur in game scoreboards, digital stopwatches, meters (such as capacitance and frequency meters), and signal averagers. What we mean by a *counter* is a device which accepts as input a series of pulses and produces as its output some representation of the number of pulses which appeared at the input.

No IC counter produces a decimal representation of the count, such as 7 or 143. Rather, they can produce only binary 0's and 1's at a set of output pins. Therefore, a set of lamps of some kind could be used to indicate what the output is, but it is usually desirable to have a display which reads in ordinary decimal form. Thus there is a need for a *decoder,* an IC which can accept as its inputs the binary 0's and 1's from a counter and can convert ("decode") that binary information into decimal form with the help of a decimal digit display device.

In this chapter we discuss counters and decoders and, in a general way, some display devices. Further details about display devices are taken up in Chapter 9.

Four-Bit Binary BCD Counters

Four-bit binary BCD[1] counters have one input for the reception of pulses and four outputs. This is shown schematically in Figure 8.1. The outputs are referred to as A, B, C, and D, with A representing the least significant bit.

[1]The BCD code is discussed in Appendix B.

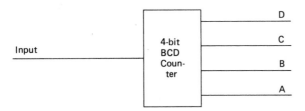

Figure 8.1 The nature of a BCD counter.

If a counter can count from 0 to 9 (decimal) in the BCD code, it is known as a four-bit *decade* BCD counter. Table 8.1 pertains to such a counter. It shows the decimal numbers which correspond to the arrival of successive pulses at the input and the states of the outputs A, B, C, and D which make up the BCD representation of each of the decimal numbers.

Table 8.1 Four-bit BCD representations of the decimal digits 0 through 9.

Decimal Number	D	C	B	A
0	0	0	0	0
1	0	0	0	1
2	0	0	1	0
3	0	0	1	1
4	0	1	0	0
5	0	1	0	1
6	0	1	1	0
7	0	1	1	1
8	1	0	0	0
9	1	1	1	1

On the arrival of the eleventh pulse [pulse number 10 (decimal)] to the counter, the output reverts to zero. Thus the counter can properly be called a *modulo-10* counter as well as a decade counter.

Since it is possible with the four outputs A, B, C, and D to represent not only 10 but 16 different numbers, the table above is a truncated version. The rest of the table is shown in Table 8.2.

Table 8.2 Continuation of Table 8.1 for decimal digits 10 through 15.

Decimal Number	BCD Output			
	D	C	B	A
10	1	0	1	0
11	1	0	1	1
12	1	1	0	0
13	1	1	0	1
14	1	1	1	0
15	1	1	1	1

There are IC counters which count from 0 (decimal) through 15 (decimal). The combination of Table 8.1 and 8.2 apply to such counters. These are modulo-16 counters. We will see in an example circuit on page 82 that we can construct modulo-n counters, where n can equal 2, 3, 5, or other numbers.

Suppose that you want a counter system which will show counts from 0 through 9999 (decimal) on a set of four decimal display units. This can be managed by using four decade counters, one associated with the units decimal digit, one with the tens digit, one with the hundreds digit, and one with the thousands digit. These must be interconnected so that the units counter activates the tens counter each time the units counter reverts from 9 to 0, and similarly for the other counters in sequence. More will be said about how this is done in due course.

Some TTL BCD Counters

Probably the most frequently used BCD decade counter is the 7490. This is a simple yet fairly versatile counter. A pin diagram for the 7490 is shown in Figure 8.2.

The A, B, C, and D outputs are at pins 12, 9, 8, and 11, respectively. In normal decade counting the input pulses go to pin 14, and pins 1 and 12 must be connected by an external jumper wire. Interesting alternatives are to omit this jumper and to bring the input pulses to pin 1 or to pin 14. In the former case the output counts from 0 to 4 (decimal) and is a modulo-5 counter. In the latter case the output counts from 0 to 1 (decimal) and is a modulo-2 counter.

The zero set and nine set pins are used to exercise still other options. If both zero set pins are brought to logic 1 states, the counter output will be set to 0. This function is often, and appropriately, referred to as "clear." If both of the nine set pins are brought to logic 1 states, the counter output will be set to binary 1001, which is decimal 9. This function is often referred to as "preset."

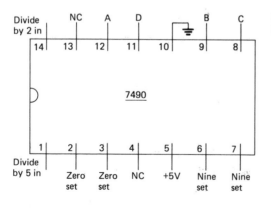

Figure 8.2   Pin diagram of the 7490 BCD decade counter.

For normal counting, at least one of pins 2 and 3 must be brought low and at least one of pins 6 and 7 must be brought low.

A more sophisticated counter is the 74193. This is a BCD counter with outputs A, B, C, and D, but it is a modulo-16 counter and it has options which the 7490 lacks. A pin diagram is shown in Figure 8.3.

The input is either at pin 5, in which case the count will proceed upward in the normal counting sequence, or at pin 4, in which case the count will proceed downward. If the clear pin (14) is at logic high, the outputs will all be 0's. In normal up-counting the output will eventually go from binary 1111 to binary 0000, and when this happens a logic change will appear at the *carry* pin. This signal can be used as an input to a second 74193 counter so that the original counter and this second counter together can count from decimal 00 to decimal 255. By further cascading a count from decimal 100 to decimal 4095 can be accomplished, or a count to larger totals. When down-counting the corresponding operation is *borrow,* the necessary signal then appearing at the borrow pin.

Finally, pins 1, 9, 10, 11, and 15 provide a nice preset function. For normal counting the *load* pin (pin 11) is high. If this pin is brought low, the outputs automatically go to the states required by whatever the inputs to pins 1, 9, 10, and 15 are. Thus to preset the counter to binary 0011, pins 9 and 10 are brought low and pins 1 and 15 are brought high, and then the load pin (11) is brought low. When the load pin is returned to high, normal counting (either up or down) is resumed, beginning at the count that was preset.

A TTL counter which is in a sense intermediate between the 7490 and the 74193 is the 74192. This device, like the 7490, is a decade counter, but like the 74193 it has up/down, clear, and preset options. It has the same pin diagram as does the 74193.

Figure 8.3   Pin diagram of the 74193 BCD counter.

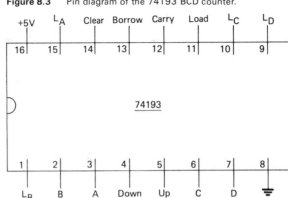

An Introduction to Decoders and Drivers

Let us suppose that we have a BCD counter in operation and want a convenient way to read the output. The output is the set of lows and highs at the A, B, C, and D output pins. One way to read the output is to put a DC voltmeter which can indicate + 5 volts at each of the counter output pins. However, this is awkward, as well as expensive. Another way is that shown in Figure 8.4. A small lamp has been connected between each output pin and the circuit ground. Any pin that is high will turn its lamp on and any pin that is low will leave its lamp off. However, this scheme will work only if the lamps are carefully selected so that the TTL counter outputs can in fact drive them on, and do so safely.

An improvement in the scheme is shown in Figure 8.5. Between each of the counter output pins and the associated lamp a "driver" has been inserted. The driver has improved current-sourcing ability over that which the counter itself has. Each driver could be a transistor which is easily turned on or off by the TTL counter outputs, but which can supply sufficient current to the lamp.

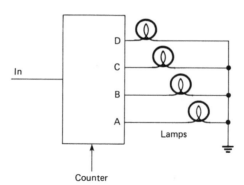

Figure 8.4 BCD counter driving indicator lamps in current-sourcing mode.

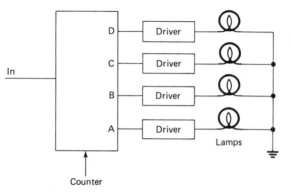

Figure 8.5 Drivers inserted into the circuit shown in Figure 8.4.

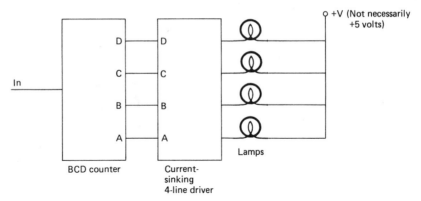

Figure 8.6 BCD counter and driver used with indicator lamps in current-sinking mode.

This is sometimes done but it requires not only the transistors but some external components (at least some resistors), and it may require a second power supply for the transistors.

A much simpler way is to use one of the TTL *drivers* which are available. These are often hex devices (six drivers to a package). If you can make a suitable match between the voltage and current demands of the lamp indicators and the properties of a driver, then the circuit of Figure 8.5 could be made practical with only two ICs (the counter and the driver) and the lamps.

The circuit in Figure 8.5 has assumed that the lamps are operated by the drivers in *current-sourcing* mode. A similar circuit which uses *current-sinking* mode is shown schematically in Figure 8.6. The distinction between current-sourcing and current-sinking is of fundamental importance. We discussed that in Chapter 5 and reference to that section should be made if necessary for understanding the basic difference between the circuits of Figure 8.5 and 8.6.

An Introduction to Seven-Segment LED Displays

A seven-segment display unit consists of seven bar-shaped elements,[2] arranged as shown in Figure 8.7. The segments can be made ON or OFF independently. The manner in which the segments are labeled a through g should be noted. The great virtue of this display is that it can show the *decimal* digits 0 through 9. It can also show a few other symbols, some of which are especially useful. (See Table 8.5.) As most everyone knows today, the seven-segment

[2]The segment may actually consist of a set of closely spaced small circular light sources but except under close inspection the viewer is not aware of this.

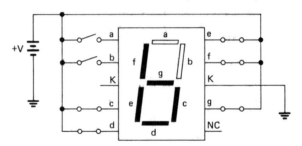

Figure 8.7   A common cathode seven-segment display operated with manual switches.

display is extremely widely used for decimal readout in calculators and elsewhere. The package shown is in DIP form and is usable in standard IC sockets and breadboards. More detailed information will be presented in Chapter 9. At present our concern is only with the general ideas involved and with the decoders and drivers needed to operate the units. We will assume that each of the segments is a light-emitting diode, or LED. (In another common type each is a liquid crystal, or LCD, segment.) Each LED segment has two wire leads coming from it and these run internally to two of the pins of the unit. One is called the *cathode* connection and this is tied to the circuit ground. The other is called the *anode* connection and this is tied to the positive terminal of some power source, which might be an output pin of some TTL device. If the anode pin is low, the LED segment is nonconducting and does not emit light. If the anode pin is brought high, the LED segment conducts and emits light. In, an actual display package, *all* of the seven cathode leads from the segments are brought out to *one*[3] pin. This form of the display unit is a *common cathode* type.

In order to solidify the rather abstract description just given, refer to Figure 8.7. In this circuit, manual operation of the switches makes possible lighting up any of the decimal digits 0 through 9 or any of some other symbols. In the drawing, the decimal number 6 has been lit. The pins labeled K are alternative common cathode pins.

Before going further we must insert a caveat. Seven-segment LED display units are made in a wide variety of shapes and sizes and with widely varying electrical properties. Many of them would be burned out if the power supply shown in Figure 8.7 furnished 5 volts. To prevent this, suitable resistors would have to be installed between each segment anode pin and the power supply positive terminal or else the voltage would have to be adjusted suitably. We

[3]Actually, two (or even three) pins of the display package are usually furnished as alternative common cathode pins for the user's convenience. They are all connected together internally and only one need be used.

postpone discussion of these matters and continue to deal with general ideas only.

Another kind of seven-segment display unit is the *common anode* type. It looks just like the unit shown in Figure 8.7 but the anode leads from the segments are all brought internally to one package pin (or perhaps two or three equivalent pins), and each segment cathode runs internally to its own package pin. The common anode pin is often marked K+ in diagrams.

To render these remarks more concrete, Figure 8.8 shows a common anode display unit in a circuit which is the counterpart of that in Figure 8.7. With a common anode unit, the cathode pin which runs internally to a segment must be brought low in order to light up the corresponding segment. In the drawing the switches have been set to cause the decimal digit 2 to be displayed.

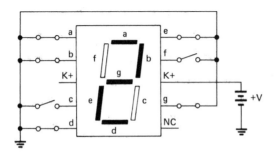

Figure 8.8   A common anode seven-segment display operated with manual switches.

BCD to Seven-Segment Decoder/Drivers

Often a BCD counter is required to drive a seven-segment display unit. This presents a problem because a BCD counter has four outputs (the A, B, C, and D output pins which we discussed before) while the seven-segment display has seven inputs (the a through g pins which we have just introduced). We can properly say that the *output code* used by the counter is different than the *input code* needed by the display unit. What is called for is some device which can accept BCD information as input and which will provide output in a code which seven-segment displays will recognize. Such a device would be a variety of *decoder*. The particular one needed in the present case is called a *BCD-to-seven-segment decoder*.

If the display unit is of the common cathode type, the decoder must obey Table 8.3. If the display unit is of the common anode type, it must obey Table 8.4.

Table 8.3 Truth table for a decoder for a common cathode seven-segment display unit.

D	C	B	A	a	b	c	d	e	f	g
0	0	0	0	1	1	1	1	1	1	0
0	0	0	1	0	1	1	0	0	0	0
0	0	1	0	1	1	0	1	1	0	1
0	0	1	1	1	1	1	1	0	0	1
0	1	0	0	0	1	1	0	0	1	1
0	1	0	1	1	0	1	1	0	1	1
0	1	1	0	0	0	1	1	1	1	1
0	1	1	1	1	1	1	0	0	0	0
1	0	0	0	1	1	1	1	1	1	1
1	0	0	1	1	1	1	0	0	1	1

Table 8.4 Truth table for a decoder for a common anode seven-segment display unit.

D	C	B	A	a	b	c	d	e	f	g
0	0	0	0	0	0	0	0	0	0	1
0	0	0	1	1	0	0	1	1	1	1
0	0	1	0	0	0	1	0	0	1	0
0	0	1	1	0	0	0	0	1	1	0
0	1	0	0	1	0	0	1	1	0	0
0	1	0	1	0	1	0	0	1	0	0
0	1	1	0	1	1	0	0	0	0	0
0	1	1	1	0	0	0	1	1	1	1
1	0	0	0	0	0	0	0	0	0	0
1	0	0	1	0	0	0	1	1	0	0

The entries in the a to g columns of either table are the logical complements of the corresponding entries in the other table, which is the only respect in which the two decoders differ logically. However, the decoder of Table 8.3 is intended for current-sourcing and that of Table 8.4 for current-sinking. The electrical properties of the two can not be expected to be identical. Provided the voltage and current demands of the display unit fall within the range of capability of the decoder, display unit pins a to g can be connected to the a to g output pins of the decoder directly. Otherwise, current-limiting resistors may be needed or the decoder may not be usable for the purpose at all.

In Figure 8.9 we show in schematic form a complete system which will accept at "Pulses in" the pulses to be counted and display the count as 0, 1, 2, 3, 4, 5, 6, 7, 8, 9, 0, 1, 2, 3, and so on. In the figure a common cathode display unit is used for the sake of illustration. If a common anode display unit was used, the decoder/driver would have to be one which obeys Tables 8.4 and one of the K+ pins would have to go to an external power supply.

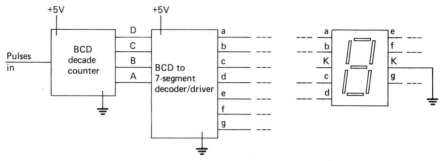

Figure 8.9   A complete counter system.

Since decoders usually have properties of drivers, they are often called *decoder/drivers* as in the figure.

The 7447 and 7448 Decoder/Drivers

The 7447 BCD to seven-segment decoder/driver is intended for use with a common anode seven-segment display unit. The pin diagram is shown in Figure 8.10.

The preceding discussions should make the meanings of the labelings of all the pins clear with the exceptions of pins 3, 4, and 5. These pins provide the 7447 with some features which have not been mentioned before.

LT means *lamp test*. If this pin (3) is brought low (by connecting it to the circuit ground) all the outputs a through g are brought low. The result of this

Figure 8.10   Pin diagram of the 7447 BCD-to-seven-segment decoder/driver.

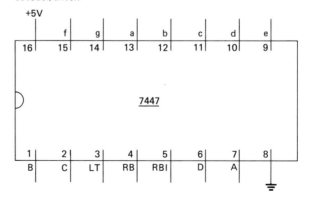

should be to cause all the seven segments of the display unit to light up. If an 8 is not seen then the display unit should be suspected of being defective.

RBI means *ripple blanking input.* If the seven-segment display unit was one of a set of (say) 12 such units, and you wanted to display the number 173 it would be undesirable to have the display panel show 000000000173. What you want is a way to suppress the leading zeroes. If the RBI pin (5) is brought low, the effect is to suppress what might otherwise be shown as a 0. (Any digit other than 0 is not affected.) When pin 5 is brought low, pin 4 also goes low. If pin 4 of the decoder is connected to pin 5 of that decoder which drives the seven-segment display unit next to the left, then the result will be that the signal which suppresses the 0 referred to will automatically suppress the 0 next to the left of that one. By extending the chain, all the leading zeroes can be suppressed. The appropriateness of the qualifier "ripple" in the name of the pin should be clear. The suppression *ripples* down the chain.

When the 7447 is used with LED displays, current limiting resistors in each of the a to g lines are required. Typically 330-ohm resistors are used.

If the seven-segment display is of the common cathode type, the 7448 must be used. While the 7447 obeys the truth table of Table 8.4, the 7448 obeys the truth table of Table 8.3. The pin diagram for the 7448 is exactly the same as that for the 7447. The 7448 is able to drive many common LED display units without the need for resistors. In much of the experimental work in IC electronics this is an advantage since the circuitry is somewhat simplified. Nevertheless, common-anode display units with 7447 decoder/drivers are most frequently used.

Table 8.5 Special characters corresponding to decimal 10 through 15 as shown on a seven-segment display unit, for the 7447 or 7448 decoder/drivers.

D	C	B	A	Symbol Shown on 7-Segment Display
1	0	1	0	
1	0	1	1	
1	1	0	0	
1	1	0	1	
1	1	1	0	
1	1	1	1	

The counter in Figure 8.9 is a decade counter. Suppose instead it was a modulo-16 counter such as the 74193 instead so that it can count from 0000 to 1111, which is decimal 0 to decimal 15, and then recycle. (Reference to Tables 8.1 and 8.2 may be helpful here.) The circuit would still work but the display unit would show not only the decimal digits 0 through 9 but, as the count continues, would go on to show the additional symbols shown in Table 8.5. The ability to do this is built into the 7448 and 7447 decoder/drivers.

It is interesting to observe that the input 1111 (decimal 15) produces no visible symbol, which is to say a blank. Since a blank may be just what is wanted in some applications, this property of the decoder may be useful.

An Example of a Circuit

As an example of a simple counter-display system we offer Figure 8.11. This is a basic stopwatch. (We have omitted the start/stop feature but suggest that you look at the pin diagram for the 74192 and invent a way to handle that feature.) The circuit shows how a two-decimal-digit display can be managed. In principle, this display could be extended to three, four, or more digits. (However, a technique called multiplexing would be used in practice if the number of displays was to be much larger than 2. Multiplexing is discussed in Chapter 11.)

Figure 8.11   Essentials of a stop-watch with two decimal digit display capacity.

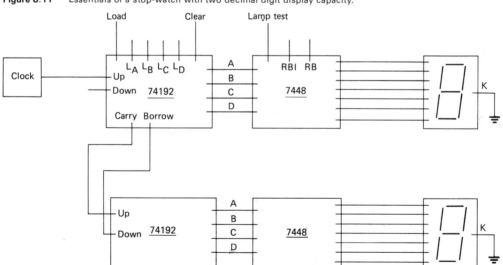

If the clock provides pulses at the rate of one per second, the display units will indicate times from 0 to 99 seconds. If the clock rate was 10 pulses per second and a decimal point[4] was used between the display units, the readout would be in tenths of seconds.

The functions of some of the pins of the uppermost 74192 and 7448 have been identified by name to suggest some of the options which are available. An obviously useful function which you would want to make use of is the clear function. Also, the ability to cause count down from some present value could be desirable.

A Divide-by-*n* Circuit

The purpose of the circuit in Figure 8.12 is to transform a stream of m pulses per second into a stream of m/n pulses per second. For example, if $m = 100$ and $n = 7$, then the output will be activated on every seventh input pulse for an output rate of 100/7 per second. The circuit shown applies when $n = 7$.

The output of the circuit remains low until the seventh pulse of a series of pulses is received at pin 14, whereupon the output goes high. This occurs because the nine-set feature is activated when the seventh pulse arrives. On the next input pulse, the output will be low again and will remain so until the next seventh pulse in sequence arrives as an input.

The circuit of Figure 8.13 provides another example. The object is now division by 6. Notice that an AND gate is used immediately before the circuit output. Verification that this circuit will work—and devising other divide-by-*n* circuits—is left to you as an instructive exercise.

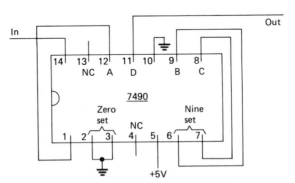

Figure 8.12   A divide-by-7 circuit.

[4]Many LED seven-segment display units have a circular "segment" to indicate a decimal point. A special DP pin is then present on the package for operating the decimal point.

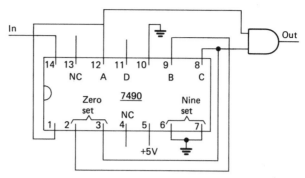

Figure 8.13   A divide-by-6 circuit.

EXPERIMENTS

1. Building a decimal digits display system

In this experiments a BCD-to-seven-segment decoder/driver is to be coupled to a seven-segment display device. Every effort has been made to make the instructions complete with respect to just what is to be done, assuming the specific apparatus referred to is used. Variations in details may be necessary, depending on what hardware you have available.

This circuit is not a useless one which is to be taken apart once the exercise has been completed. On the contrary, it is highly useful and will be used in later experiments as a display unit. For that reason the circuit should be built compactly up in one corner of a breadboard where it will be available later, while leaving room for the circuitry which other experiments will need.

Our instructions assume use of a 7448 TTL decoder/driver and of a MAN-4[5] seven-segment LED display device or an equivalent common cathode unit. If a common anode display is to be used, the 7447 decoder/driver should be used instead of the 7448 and a resistor of about 330 ohms should be installed in the line to each cathode segment of the display unit. The pin identifications for the 7448 are given in Figure 8.10. (This figure shows a 7447, but the pin identifications apply equally to the 7448.) The pin connections of the MAN-4 are shown in Figure 8.14.

Install the display device and the decoder next to each other in one corner of a breadboard as in Figure 8.15 or at one end of a breadboard *strip,* depending upon the format of your breadboard.

---

[5]The MAN-4 is referred to not because it has any special virtue as a common cathode unit, but because it is so typical. Any of perhaps a dozen others will serve just as well.

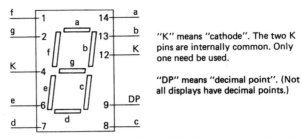

"K" means "cathode". The two K pins are internally common. Only one need be used.

"DP" means "decimal point". (Not all displays have decimal points.)

Figure 8.14 Pin diagram of the MAN-4 seven-segment display unit.

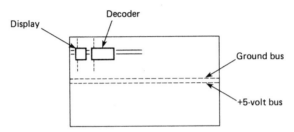

Figure 8.15 Locating the decoder and display unit on a breadboard to leave room for other circuitry to be added later.

Interconnect output pin a of the decoder and the pin a of the display with a length of wire. Proceed similarly with the b through g connections. Wire one or the other of the K (cathode) pins of the display to the board ground bus. (Only one of the two K pins need be used.) Wire the ground pin (pin 8) of the decoder to the board ground bus. Wire the +5-volt pin (pin 16) of the decoder to the other board bus. Wire the ground terminal of the power supply to the board ground bus and the +5-volt terminal of the power supply to the board *high* bus.

The pins of the display unit have now all been connected but the following pins on the decoder remain unconnected: A, B, C, D, LT, RB, and RBI. Our concern now is with A, B, C, and D. We will do something with the other three presently.

When the decoder is suitably commanded, it will cause a corresponding symbol to light up on the display. For example, if you tell the decoder in a way it understands that you want a 7, the decoder will see to it that the display acts accordingly. The question is how you inform the decoder of what you want.

This is done by connecting each of the pins, A, B, C, and D of the decoder either to the circuit ground or to the +5-volt supply. The symbol which is to be displayed must be given to the decoder in binary form. How BCD inputs are converted into visible characters is shown in Table 8.6. In the table a 1 means "connect to +5 volts" and an 0 means "connect to ground."

Table 8.6 Complete table of characters shown by a seven-segment display unit driven by a 7447 or 7448.

Decoder Input Pins				Corresponding Symbol Shown on Display Unit
D	C	B	A	
0	0	0	0	0
0	0	0	1	1
0	0	1	0	2
0	0	1	1	3
0	1	0	0	4
0	1	0	1	5
0	1	1	0	6
0	1	1	1	7
1	0	0	0	8
1	0	0	1	9
1	0	1	0	See (a) at right
1	0	1	1	(b)
1	1	0	0	(c)
1	1	0	1	(d)
1	1	1	0	(e)
1	1	1	1	(f)

(a) (b) (c) (d) (e) (f)

Now the circuit can be tested. Choose some symbol which you would like to have displayed and connect the A, B, C, and D pins of the decoder correspondingly. This can be done simply by using lengths of wire, or if you have suitable switches they can be used to make the connections at will. Turn on the power to the circuit. Try lighting up all the symbols.

A peculiarity of TTL ICs is that an input pin acts the same whether it is connected to + 5 volts or is left unconnected. This can be verified by leaving 1 inputs to the pins dangling (unconnected) instead of connecting them to +5 volts. (See the Important Special Properties of TTL ICs section of Chapter 3.)

Pin 3 is labelled LT in Figure 8.10. The name means "lamp test." Its purpose is to provide a simple way to find out if all of the seven segments of the display are working. Normally LT is left unconnected or is connected to +5 volts. If it is brought low by running a wire from pin 3 to the circuit ground, all of the segments of the display should come on and you should see an 8.

Pins 4 and 5 of the decoder are RB and RBI. The functions of these pins were discussed in Introduction to Seven-Segment Displays section of this chapter.

2. Building a complete counter system

Our purpose here is to design and build a digital IC system which involves many of the basic ideas of modern electronics. The circuit is to fulfull the following specifications: (1) There is to be a system input at which pulses can be applied. These can be from any source as long as they are TTL-compatible. In this experiment the inputs can be applied manually by use of a debounced switch or they can be taken from the 555 timer clock. (2) The count is to be displayed on a seven-segment display unit. (3) The system is to count *up* or *down* on command. (4) The count can at any moment be reset to zero. (5) There is to be provision for presetting a count when that is desired.

We will assume that a 74193 counter is used. We suggest that as a follow-up experiment the popular 7490 counter be used in order to become familiar with it. (See experiment 3.) However, the 74193 has features which are lacking in the 7490 and it is the desirable choice for this experiment.

Examine the pin diagram for the 74193 and determine which pins are for the ground and +5-volt power connections, which accept the input pulses, which provide the output data, and so on. The 74193 is to drive the 7448 decoder which in turn drives the seven-segment display, as described in the first section of this chapter. Determine how the 74193 outputs are to be connected to the inputs of the decoder. Then draw a diagram which shows the complete circuit to be constructed.

When the circuit design is complete, wire up the circuit. When you feel that no mistakes are present, apply power to the circuit. In order to supply input pulses to the counter, connect a wire to an input pin of the counter and alternately push the other end of the wire into the ground bus of the breadboard and pull that end out. You should expect that pulsing the counter in this way will give erratic results due to switch bounce. Then install a debounced switch as the pulse source and observe the improved operation.

After studying manual pulsing of the counter system, convert to *continuous* operation by hooking up the output pin (pin 3) of the 555 timer to the input pin of the 74193.

Try both the count up and count down input pins of the counter.

Students designing and building this circuit sometimes have a counter which will not work because they have not taken the clear and load pins of the 74193 seriously enough. For normal counting the load pin must be high and the clear pin must be low. To leave the load pin unconnected to anything is all right, but the clear pin *must* be connected to the circuit ground in order to bring it low.

When the system counts up and down properly, you should be sure that you can preset data. This requires inputting the preset data at pins 1, 9, 10, and 15, and using the load pin properly.

3. A system to count higher

In a group laboratory situation there will eventually be several operating single-stage counters, and it is worthwhile to hook two or more of them together to make a single composite counter which can count higher than one alone can. If you are working alone, you could make a second counter circuit and then couple the two counters together. The carry pins of the 74193 are used for *cascading* for up-counting, the borrow pins for down-counting.

4. Using the 7490 counter

In the preceding experiments the 74193 was chosen as the counter because it has options which make it very instructive. Today the 7490 decade counter is no doubt the most frequently used counter and you owe it to yourself to get acquainted with it. An almost *must* experiment is to design and build a counter system using the 7490 and the decoder/driver-display unit. This is not entirely trivial for the newcomer to the 7490 because of some intricacies. These include the need to connect two input pins together and to use the zero-set and nine-set pins correctly.

5. Gating the 74193 counter system

Experiment 2 has to do with a counter system based on a 74193 counter. The 74193 is not a decade counter. A very instructive problem to tackle is this: Determine how the 74193 counter system can be converted into a decade counter by gating. *Gating* means that some basic logic gates are to be used. A hint: As the counter procedes from (decimal) 0 through 1, 2, 3, and so on, the D bit of the BCD representation of the count never changes from 0 to 1 until 8 (decimal) is reached and 9 (decimal) and 10 (decimal) have the unique properties that for the first times in sequential counting D becomes 1 with either A or B also becoming 1.

6. Divide-by-n

In Experiment 5 the idea was to use gates to convert a modulo-16 counter into a modulo-10 (decade) counter. A generalization of this idea would be to convert a modulo-16 counter into a modulo-n counter, with n in the range from 1 to 10. A counter so modified is called a divide-by-n counter. Divide-by-n is a technique with many applications. An instructive experiment is to choose a value for n such as 9 (for example), design a gating system to make a counter divide by that value of n, and build the circuit to verify that the design was correct. (The Divide-by-n section deals with the cases where $n = 6$ and $n = 7$.)

The Most Common
Display Devices

The Discrete LED

The *light-emitting diode*, or *LED*, is a special kind of lamp. It comes in various shapes, two of which are shown in Figure 9.1. It makes a good pilot lamp to show whether the power to some instrumentation is on or off and it can be very useful as an indicator to show whether the logic state of a pin or an IC is high or low.

The usefulness of the LED as a logic state indicator alone justifies a discussion of it but, in addition, since each of the segments of the common LED seven-segment display unit *is* a light-emitting diode, your understanding of the seven-segment display is enhanced by understanding the discrete LED.

A discussion of the LED must begin with the *semiconductor diode.* This consists of P-type semiconductor material joined to N-type semiconductor material. The materials are silicon (most often) or germanium. This simple structure is represented schematically in Figures 9.2, (a) and (b). In Figure 9.2(a) the P-type end is connected to the negative terminal of the battery. This constitutes *reverse bias,* and the result is that the diode is in a nonconducting state. Simply put, it is *off.* In Figure 9.2(b) the battery polarity has been reversed. This is *forward bias,* and the diode conducts current. It is *on.*

In Figure 9.2(c) and (d) the same circuits are redrawn, this time making use of the conventional symbol for a diode. In (c) the diode is off and in (d) it is on.

The LED is a diode and therefore if it is reverse biased as in Figure 9.2(c) it is off, and if it is forward biased as in (d) it is on. But, unlike ordinary diodes,

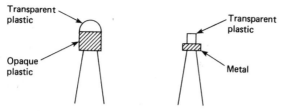

Transparent
plastic

Opaque
plastic

Transparent
plastic

Metal

Figure 9.1 Two forms of LEDs.

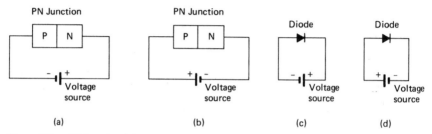

(a)   (b)   (c)   (d)

Figure 9.2 PN junction (a) reserve biased; and (b) forward biased. The same junction, using the conventional diode symbol: (c) reverse biased; and (d) forward biased.

the LED emits light when it is on. As indicated in Figure 9.1, the LED package has a transparent portion so that the light emitted can be seen. In Figure 9.3 we show the conventional symbol for an LED in a circuit which biases the LED on. If the battery polarity was reversed, the LED would not emit light.

We will discuss this circuit in more detail in order to bring out some important practical points about it. One is that when an LED is biased on there is a voltage drop across it. This *forward drop* amounts to about 1.6 volts for many LEDs. This is not an IR drop, which would vary with the current, but rather it is associated with the nature of the interface between the parts of the diode and is a constant. Another important point is that LEDs differ with respect to the amount of current they can carry safely when they are on. For one the current may be 20 milliamperes per segment and for another only 10 milliamperes or less.

It is the presence of the forward drop in the LED and the limitation on the current that dictate the presence of the resistor R in the circuit. Suppose the

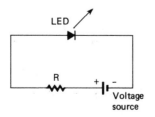

LED

R

Voltage
source

Figure 9.3 Circuit for operating an LED.

LED in the circuit cannot tolerate a current above 10 milliamperes. The voltage across R will be (5 − 1.6) volts, or 3.4 volts. Applying Ohm's law, we find that R must be equal to 340 ohms. A smaller value for R may increase the intensity of the light somewhat, but at the risk of death for the LED. A larger value will reduce the light intensity.

One kind of LED is so commonly used in pilot lamps and in the seven-segment displays of calculators that nearly everyone has seen the characteristic reddish color it emits. This kind of LED owes its color to the substance of which it is made: gallium arsenide phosphide (GaAsP). The light consists of wavelengths principally at about 670 nanometers. LEDs which emit yellow light or orange light or green light are also available, and LEDs which emit in the infrared region can be obtained. These IRLEDs are popular with experimenters but they have no particular usefulness in digital IC circuits.

More About LED Seven-Segment Displays

In Chapter 8 there was a substantial discussion of how LED seven-segment displays are used with suitable decoder/drivers. We saw that each segment of the display unit is a light-emitting diode. The discussion in the first section of this chapter should help you understand LED seven-segment displays. In this section we offer a summary of other properties of LED seven-segment units.

There are many kinds of LED seven-segment units on the market. They differ in physical size, brightness, and current demand. They can be had singly or in packages of 2 to 12 or more. They may be able to display a decimal point, a colon (useful in digital clocks), or other symbols. They are, of course, either of the common cathode or of the common anode types. The problem for you as a user is making a choice suitable for your purpose.

Table 9.1 briefly describes some LED seven-segment displays to give you an idea of the range of characteristics available.

Table 9.1 Comparison of some characteristics of some LED seven-segment display units.

Type	Character height (centimeters)	Current per segment (milliamperes)	Common anode or common cathode
MAN 3	0.30	10	Common cathode
MAN1	0.69	30	Common anode
FND10/a	0.69	10	Common cathode
SSL 190	0.69	30	Common anode
747	1.52	25	Common anode
FND507	1.27	20	Common anode

Any diode can be ruined by excessive reverse bias. For most diodes the manufacturer states the maximum reverse bias allowable. For LEDs the maximum reverse bias is not as definite and may not even be mentioned. However, there is no danger of excessive reverse bias in TTL IC circuitry, provided you install a discrete LED or a seven-segment unit with the correct polarity.

Matrix Dot Displays

An *alphanumeric* display is one which can show numbers and letters. Usually it can also show other symbols such as ?, *, and +. The seven-segment display is definitely not an alphanumeric unit. A display device which has as its light-emitting elements dots arranged in an array of columns and rows is capable of exhibiting a wide variety of symbols if the numbers of columns and rows are large enough. One such display device is the *5-by-7 matrix* unit. Figure 9.4 shows the arrangement of dot LEDs in it. As an example, the unit is showing the plus symbol (+).

It is quite possible to use a set of 35 switches in the manner of the circuit in Figure 8.8 to activate selected dots. Thirty-five is a large number of switches. Furthermore, if a substantial number of 5-by-7 units are to be connected to their IC drivers in some digital system, the number of drivers and interconnecting wires becomes formidable. In practice a technique called *strobing* or *scanning* is used. The elements in one row are activated simultaneously, then the elements in the next row and so on, as in Figure 9.5(a). This is *vertical scanning.* Alternatively, horizontal scanning can be used as in Figure 9.5(b). The scanning is carried out so fast that the viewer of the display is not aware of any flicker. This technique requires careful clock timing and special decoders. This is a usage of a method known as *multiplexing,* a subject to be discussed later in this book.

There are also 4-by-7 matrix display units. A 4-by-7 can display more characters than can a seven-segment unit but it is more restricted in its capabilities than is a 5-by-7.

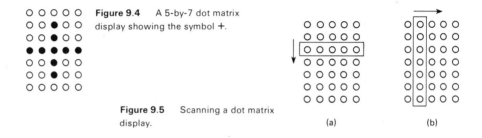

Figure 9.4   A 5-by-7 dot matrix display showing the symbol +.

Figure 9.5   Scanning a dot matrix display.

(a)   (b)

Linear LED Arrays

There are now available linear (side-by-side) arrays of discrete LED indicators. The array may contain from a few to as many as 50 LEDs in a line. Special drivers are also on the market.

The usual applications for these are of the bar graph kind. For example, some sensor in a tank of a fluid may signal a digital system with the ultimate result that one of a set of LEDs in a linear array goes on. You can, at a glance, read an approximate measure of the contents of the container. For another example, a voltage probe applied to a point in some circuitry may cause the appropriate LED in a linear array to go on, enabling the user to get a rough but easily readable idea of the voltage.

Liquid Crystal Displays

Liquid crystal displays, or *LCDs,* are much like LED displays in some ways but they differ in important respects. One is that LCDs scatter light while LEDs generate light.

A liquid crystal is an organic chemical compound for which the state of aggregation varies according to the temperature. Below a certain temperature the compound is a crystalline solid. If the temperature is above a certain higher temperature, the compound is a clear liquid. At temperatures between those two critical temperatures, the compound has some of the properties of a solid and some properties of a liquid. *That* region is the liquid crystal region. When it is in that region, the substance has a murky yellowish appearance.

When in the liquid crystal state, the compound is sensitive to an applied electrical field. The molecules of the liquid crystal are long and each has a positive and a negative end. We say they are *electric dipoles.* In the presence of an electric field, the molecules tend to line up with their axes along the direction of the field.

When a liquid crystal substance is used in an LCD unit, it is the influence of the aligned molecules in an electric field on *light* passing through the substance that is of importance. The structure of a *field-effect* liquid crystal cell is illustrated in Figure 9.6.

The liquid crystal substance is sandwiched between two glass plates. The front plate bears the seven-segment pattern, and possibly other symbols such as a colon which would be used in digital clock display. These are usually made of indium oxide, a transparent substance which is electrically conducting. Each of the segments has a metal lead to an external pin on the unit. On the back plate there is an electrode, shaped to cover the segments on the front plate. A metal lead goes from this electrode to an external pin on

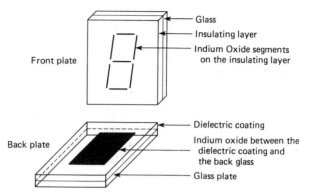

Front plate

- Glass
- Insulating layer
- Indium Oxide segments on the insulating layer

Back plate

- Dielectric coating
- Indium oxide between the dielectric coating and the back glass
- Glass plate

Figure 9.6   Internal structure of a field-effect liquid crystal cell.

the unit. Insulating layers separate the conducting layers from each other and from the glass plates.

Finally, polarizing sheets and a reflector sheet are added, as in Figure 9.7. In what is known as a *reflective LCD*, the polarizer sheets are crossed. This means that their axes of transmission are at 90 degrees to each other. With light incident as shown in the figure, the first polarizer transmits only the vertically polarized component of the light. The liquid crystal material rotates the lane of polarization so that the light is passed by the horizontal polarizer. The light then falls on the reflector, returns back through the cell, again rotating sufficiently to get through the vertical polarizer, and finally passes out to the region from which it came.

Under those conditions you would see the face of the unit as rather uniform in appearance. But if an electric field is applied to some of the segments on the front plate,the molecules of the liquid crystal will align themselves with the field with the result that the rotations which were described above do not occur. The light which is not rotated properly does not get

Figure 9.7   Functioning of a reflective LCD display unit.

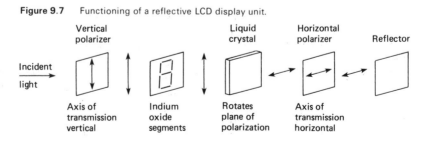

to the reflector plate and does not exit again to the left. The result is that the chosen segments appear dark against the otherwise light background.

There is also a *transmissive LCD*. The structure is essentially the same as that shown in Figure 9.7, but the reflector plate is lacking so that incident light from the left is polarized by the first polarizer plate, rotated by the liquid crystal material, passes through the second polarizer plate, and *exits to the right*. An eye looking from that side would see a general faint illumination.

If a field is applied to some chosen segments, the corresponding parts of the light do not have their polarization directions rotated properly and these parts do not get through the unit. The eye would see them as dark against a lighter background.

Both of these kinds of LCD displays are known as *field-effect LCDs*. In addition to these, there are *dynamic-scattering LCDs,* both reflective and trans-missive, but you must refer to more specialized literature for a discussion of these.

Some Technical Characteristics of LCDs

Some characteristics of LCD displays are of considerable practical importance. One is that while LED displays *generate* their light and so are usable in the dark or can compete with ambient lighting which is not too bright, field-effect LCDs cannot be used in the dark unless there is some light source external to the display unit itself. Given the light source, field-effect LCDs may be more readable than LEDs.

One of the most striking advantages of LCDs over LEDs is their much smaller power consumption. The internal resistance in an LCD may be of the order of 10^9 ohms. If the applied voltage is a few volts, the current will be of the order of a few billionths of an ampere, even for a large display. A similar LED display might require hundreds of milliamperes and so perhaps 100,000,000 times as much current. It is no wonder that battery-operated pocket cal-culators often use LCD displays, even though the relative brilliance of LED displays is desirable.

LEDs are generally better choices with respect to temperature condi-tions, LCDs become sluggish at low temperatures, and high temperatures [say 55°C (130°F)] are not safe for them.

Readers who are interested in experimenting with LCDs should be aware of a problem which may arise if a TTL chip is used to drive a LCD. In the first section of Chapter 5, we pointed out that when the output pin of a TTL device is in its logic low state, the voltage at that pin may be several tenths of a volt above zero. This voltage could turn on an LCD segment when that is not the intention. This can be avoided by special techniques using TTL drivers. The more usual way to avoid the difficulty is to use a CMOS driver rather

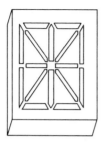

Figure 9.8   Form of an alphanumeric display unit.

than a TTL driver. The CMOS driver will give zero voltage at an output pin when the pin is in its low state. (CMOS is discussed in Chapter 15.)

To conclude this section we point out that LCD units are available with more than seven segments per unit. A useful one is shown in Figure 9.8. This is an alphanumeric display which can show all the decimal digits, all capital letters and some special symbols such as + and −.

The Nixie Tube

The Nixie tube is a gas discharge device and it is quite different than the display devices we have described so far. Nixie tubes are used in large numbers in measuring instruments, copying machines, and elsewhere. The Nixie was developed by the Burroughs Corporation in the 1950s and the name is a registered trademark.

The Nixie consists of ten metal pieces which are shaped into the forms of the decimal digits. This is illustrated in Figure 9.9.[1] The metal pieces are insulated from each other and each can be used as a cathode. There is a single anode in the tube. This is a piece of fine metal mesh in the form of a half cylinder. The parts are packaged in a glass bulb in which there is a small amount of mercury vapor.

If a suitable voltage is applied between the anode and any one of the cathodes, the vapor breaks down and light is emitted in the region near the cathode. A viewing eye sees the corresponding numeral glow with a red-orange appearance and the unlit numerals are not noticed.

A disadvantage of the Nixie tube is that a voltage much larger than that used in TTL or CMOS circuits is required. Typically +175 volts DC is used with a current-limiting resistor of 15,000 ohms in the line. The current needed is about 7 milliamperes. Special Nixie transformers are available on the market. A special decoder/driver is also needed. One such unit is the TTL

[1]From "Electronic Numbers", by Alan Sobel. Copyright © 1973 by *Scientific American,* Inc., page 66. Used with permission of W. H. Freeman and Company, publishers of "Scientific American." All rights reserved.

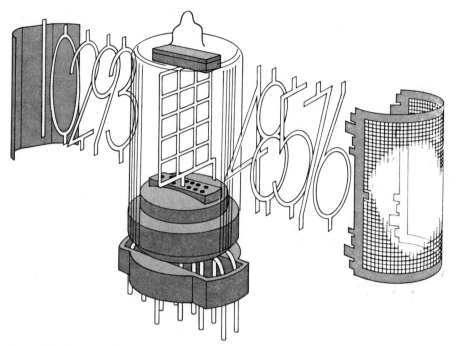

Figure 9.9　Structure of a Nixie tube.

74141 BCD-to-Nixie driver. This device accepts as inputs the A, B, C, and D binary digits from a preceding device and converts the information into a form suitable for illuminating a Nixie element. Therefore, this device is also properly called a BCD to one-of-ten decoder/driver.

　　The pin diagram for a 74141 is shown in Figure 9.10. The underscored

Figure 9.10　Pin diagram of the 74141 Nixie decoder/driver.

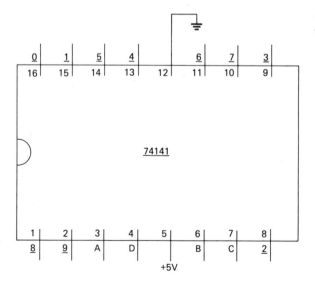

96

numbers correspond to the cathodes of the Nixie tube. The outputs can tolerate up to 60 volts and can sink 7 milliamperes when low.

It is possible to have the BCD inputs at pins 3, 4, 6, and 7 represent a decimal number larger than 9. In any such case, the outputs *all* go to an open-circuit condition and no numeral is shown by the Nixie.

EXPERIMENTS—SOME COMMENTS

This chapter has as its purpose furnishing more information about display devices than was given in Chapter 8. Refer back to the experiments described at the end of Chapter 8. The first of those gives highly detailed instructions for constructing a system in which there is a decoder and a seven-segment display unit, and for learning how to operate the system to cause any of the possible characters to appear on the display unit. The display unit was of the LED type. You may wish to experiment with an LCD seven-segment display unit as well. In that case, you need a suitable decoder/driver, as has been pointed out in this chapter.[2] Also, the inherently attractive Nixie tube may interest the reader. Unfortunately, Nixie tubes are not as easily found for sale by dealers as are LEDs and LCDs, generally speaking, and you also need a suitable transformer and decoder/driver. For the purpose of learning the basic ideas associated with decoder/drivers and displays, Experiment 1 of Chapter 8 is very effective, and the hardware needed is easily obtained.

[2]To drive an LCD display with a CMOS decoder, a square wave voltage is used. DC drive of LCDs shortens the life of the displays.

10

Flip-Flops

We saw in Chapter 7 that there are three varieties of multivibrators: the astable or free-running; the monostable or one-shot; and the bistable or flip-flop. The enormously important subject of flip-flops will occupy us in this chapter. As we will see, there are three kinds of flip-flops. They are known as the *RS,* the *D,* and the *JK.* They will be introduced in that order and several examples of the use of each kind will be given.

In Chapter 3, the distinction between combinatorial logic and sequential logic was explained. A sequential logic digital circuit produces output states which depend not only on the states of the input pins (as is true for a combinatorial logic circuit) but also on the states present in a preceding stage in the operation of the circuit. This requires that devices operating in sequential logic have some ability to remember what happened earlier. Indeed, as we will see, flip-flops are in part memory devices, although this feature is not their only virtue. As we explore the uses of flip-flops in this and later chapters, we will see that the combination of gates and flip-flops makes possible versatile and sophisticated digital circuits.

Yet another distinction between two kinds of digital logic will make its appearance in this chapter. Circuits of one kind respond to changes at the inputs as soon as the latter appear. Circuits of the other kind do not react immediately on receipt of input signals, but instead wait until an activating *clock* signal arrives at a special input, whereupon the output states change. The former kind of circuit is said to be *direct,* and the latter is said to be *clocked.* Sometimes the direct kind is called *asynchronous,* and the clocked kind *synchronous.* However, these two words are often applied to clocked logic. When that is so, what is meant is that in synchronous clocking the activating pulses arrive periodically (with a fixed frequency) while in asynchronous clocking the pulses arrives irregularly.

The RS Flip-Flop

Let us start with the simple circuit of Figure 10.1(a). This is made up of only two NOR gates in a simple configuration. The truth table for this circuit is shown in Figure 10.1(b).

   In the first line of the table, S = 0 and R = 1. This results in output Q being 0. We say that Q has been *reset* to O. The output $\overline{Q}$ is the complement of Q and so $\overline{Q}$ = 1. In the second line, S = 1 and R = 0. This results in Q = 1, and Q is said to be *set* to 1. (Of course, $\overline{Q}$ = 0 now.) This *reset-set* feature in these two lines of the table gives rise to the RS in the name of the circuit, which is *RS flip-flop*. Why is it appropriate to call the circuit a "flip-flop"? It is because by putting S and R in their 0, 1 states and in their 1, 0 states alternately, the output Q can be made to flip and flop back and forth between 0 and 1. (Of course, $\overline{Q}$ also flip-flops.) Each of the output states for Q and $\overline{Q}$ is stable. Hence the name *bistable multivibrator*.

   The last two lines of the table of Figure 10.1(b) are of special interest, the third line because it is most unusual and the fourth because it is of great practical value. In order to explain the peculiar nature of the third line, we start by pointing out that if each of S and R is set equal to 1, the outputs should be Q = 0 and $\overline{Q}$ = 0, even though in this case Q and $\overline{Q}$ are not complementary to each other. Then we suppose that after S and R are set equal to 1, each of S and R are set equal to 0. An inspection of the circuit will convince you that it is not possible to predice unambiguously what the outputs Q and $\overline{Q}$ will be. There is no preferred state. You might say that to avoid this ambiguity the circuit should never be used with S = R = 1 followed by S = R = 0 and that that can be done by avoiding S = R = 0. However, we are about to see that to have

Figure 10.1   (a) RS flip-flop made with two NOR gates; (b) its truth table.

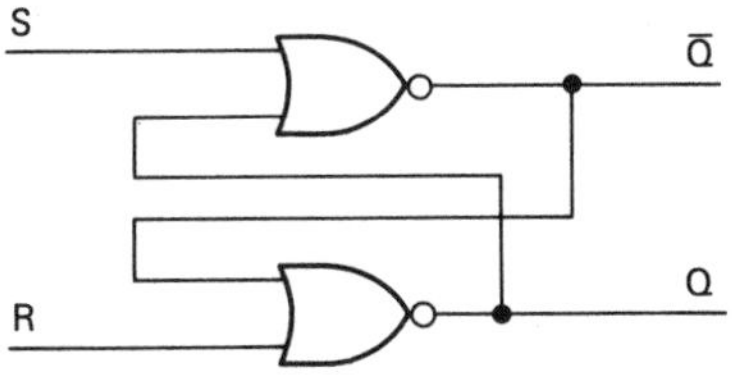

S	R	Q	$\overline{Q}$	
0	1	0	1	Reset
1	0	1	0	Set
1	1	0*	0*	(See text)
0	0	(0	1)	or (1 0)

(a)   (b)

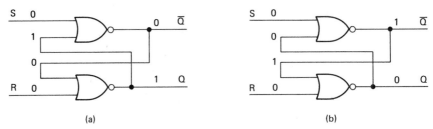

Figure 10.2 Analysis of the RS flip-flip of Figure 10.1.

$S = R = 0$ can be very important. It follows that the rule to adopt is to avoid $S = R = 1$. In the table the Q and $\overline{Q}$ outputs in the third line are marked with asterisks (*) to stress this peculiarity.

To begin our discussion of the fourth line of the truth table, refer to Figure 10.2. In the two parts, (a) and (b), of this figure $S = R = 0$. You can convince yourself that the rest of the states shown in part (a) are just as defensible as are those shown in part (b). What determines whether the outputs will be $Q = 0$ and $\overline{Q} = 1$, or $Q = 1$ and $\overline{Q} = 0$? This seems to be like being balanced on the top of a mountain so precariously that a slight push either way causes you to fall off, but definitely in one direction or the other. That is, some slight imbalance in the circuit, such as must inevitably be present, will cause the circuit to choose one or the other of the two conditions.

However, imagine that the inputs were $S = 0$ and $R = 1$ *before* the inputs became $S = R = 0$. The original output states $Q = 0$ and $\overline{Q} = 1$ were established initially, and when S and R go to 0 these output states will remain as preferred states. In a similar way, if $S = 1$, $R = 0$, $Q = 1$, and $\overline{Q} = 0$ initially, and then S and R go to 0, the output states $Q = 1$ and $\overline{Q} = 0$ will remain unchanged.

When the inputs made are $S = R = 0$, the RS flip-flop *remembers* what the last output states were. It would be hard to exaggerate the usefulness of this. In the RS flip-flop we have a basic memory device.

The D Flip-Flop

The D in the name *D flip-flop* comes from the word "Data." The significance of this will be made more clear presently but the main point is that the D flip-flop will, on receipt of a command signal at one input, "look" at whatever *data* (0 or 1) may be present at another input, and do something with the data. The command signal input is often called *Clock*.

Following these preliminary remarks, we go directly to the heart of the matter by presenting the schematic in Figure 10.3(a) and the state table in

Figure 10.3(b). There are *four* inputs to the flip-flop. They are called clock, D, S, and R. There are two inputs, Q and $\overline{Q}$, which are complementary.

Notice that the last three lines of the state table are independent of the states of the D and Clock inputs. We can set or reset Q by using S and R inputs. The D flip-flop contains an RS flip-flop, in short. Furthermore, $S = R = 1$ is to be avoided, as was true with the RS.

When $S = R = 0$ we have a divergence from the behavior of the RS flip-flop because now the nature of what happens at the Clock input becomes important. The notation $0 \rightarrow 1$ means that the state of the Clock input is initially low and goes high and $1 \rightarrow 0$ means that the state of the Clock input is initially high and goes low.

As a matter of terminology, let us call the transition $0 \rightarrow 1$ the *leading edge* of a pulse and $1 \rightarrow 0$ the *falling edge* or *trailing edge* of a pulse.

Before going further with this discussion of Figure 10.3(b), we must break off for some essential comments. A D flip-flop which obeys the state table shown is of a particular kind which is said to be *leading-edge triggered.* The action specified in the first two lines of the table occurs when the clock pulse goes from low to high. Other flip-flops react when the clock pulse goes from high to low. They are said to be *falling-edge triggered.* There are also *level-sensitive* flip-flops. In Figure 10.3(b) and the accompanying discussion we have assumed for the sake of definiteness that the flip-flop is leading-edge triggered. If it were falling-edge triggered, the clock transitions in the table would be reversed. (The terminology "positive edge" for leading edge and "negative edge" for falling edge is also used.)

We resume our discussion of Figure 10.3(b) by stating simply what happens when the conditions in the first two lines occur. On the leading edge

Figure 10.3   (a) Schematic of a D flip-flop; (b) its truth table.

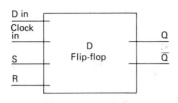

Clock	D	S	R	Q	$\overline{Q}$
$0 \rightarrow 1$	0	0	0	0	1
$0 \rightarrow 1$	1	0	0	1	0
$1 \rightarrow 0$	X	0	0	No change	
X	X	0	1	0	1
X	X	1	0	1	0
X	X	1	1	Not used	

"X" means "immaterial" or "can be either 0 or 1"

(a)   (b)

of a clock pulse when S = R = 0, the flip-flop responds to a 0 at the data (D) input by making Q be 0 and $\overline{Q}$ be 1. On the leading edge of a clock pulse when S = R = 0, the flip-flop responds to a 1 at the D input by making Q be 1 and $\overline{Q}$ be 0. With justification we can say that if S = R = 0 and the clock input goes high, the flip-flop *samples* the D input line.

According to the third line in the state table, a falling edge of the clock signal causes the flip-flop to hold whatever output states Q and $\overline{Q}$ were brought into existence previously. (This is so only if S = R = 0 also.) It would be appropriate to say that when the third line is achieved, the flip-flop *remembers* what the last outputs were. We say that when the third line is achieved the states of the outputs are *latched.*

The operation of the flip-flop when S = R = 0 can also be described as *sample-and-hold,* although this is not common terminology. This sample-and-hold ability of the D flip-flop is its most striking characteristic.

The D Flip-Flop as a Toggle

We direct your attention to an interesting way in which a D flip-flop can be used. The circuit is shown in Figure 10.4.

The $\overline{Q}$ output is connected back to the D input. Whenever the clock input changes from 0 to 1 the flip-flop looks at the D input. Suppose it sees a 0 there. Then it puts Q at 0 and $\overline{Q}$ at 1. On the next clock pulse, it sees a 1 at the D input and its put Q at 1 and $\overline{Q}$ at 0. It *toggles* the state of Q (and of $\overline{Q}$ as well) on *each* upward Clock transition. On each downward Clock transition it holds the outputs it had just previously.

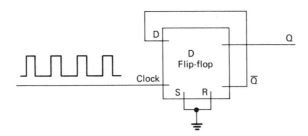

Figure 10.4   A toggle circuit based on a D flip-flop.

The JK Flip-Flop

The *JK flip-flop* does not have just four inputs (as does the D flip-flop), but five. For that reason it is more complex than is the D flip-flop. The operation of the JK is best understood by analyzing its state table. This is given in Figure

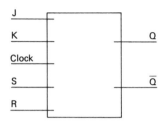

	Inputs				Outputs	
Clock	J	K	S	R	Q	Q̄
0→1	1	1	0	0	Toggle	
0→1	1	0	0	0	1	0
0→1	0	1	0	0	0	1
0→1	0	0	0	0	No change	
1→0	X	X	0	0	No change	
X	X	X	1	0	1	0
X	X	X	0	1	0	1
X	X	X	1	1	Not used	

"X" means "immaterial" or "either 0 or 1"

(a)   (b)

Figure 10.5 (a) Schematic of a JK flip-flop; (b) its truth table.

10.5(b). A schematic representation of a JK is given in Figure 10.5(a). Again we assume leading-edge triggering for definiteness.

The last three lines of the state table are like the corresponding lines of the D flip-flop state table. The JK flip-flop, in short, contains a D flip-flop. The lines in which Clock transitions matter are those to which we must give special attention.

With $S = R = 0$ and with $J = K = 1$, an upward Clock transition makes the output Q (and the output Q̄) toggle. If J and K are 0, 1 or 1, 0 an upward Clock transition makes Q agree with J (and Q̄ agree with K), provided still that $S = R = 0$. This is the set-reset function, but in response to J and K instead of to S and R.

The fourth line in the table provides a memory capability for the flip-flop. With $S = R = 0$ still, if J and K are brought low and an upward clock transition occurs, the flip-flop remembers what it last had at its outputs. Finally, any downward transition of Clock is ignored. The outputs remain as they were.

An Important Use for JK Flip-Flops: A Binary Counter

Figure 10.6 shows a series of four JKs. A source of rectangular pulses off to the left (not shown explicitly) provides input to the clock input of FF1. The Q̄ output of FF1 drives the input of FF2, and the Q output of FF1 becomes one of four outputs for the whole system. There are similar connections for the remaining flip-flops, except that FF4, being the last in the chain, has no load for its Q̄ output. Each flip-flop is assumed to have $J = K = 1$ and $S = R = 0$.

This circuit is in fact a binary counter. As the pulses from the source off to the left feed into the system, outputs A, B, C, and D of the system will count the pulses in binary form, modulo-16. By modulo-16 we mean that the

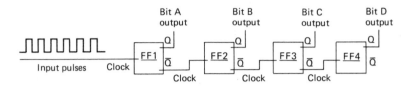

For each flip-flop, J = K = 1 and S = R = 0

Figure 10.6   Four JK flip-flops used as a four-bit BCD counter.

smallest output will be 0000 and the largest 1111 or decimal 0 at the least and decimal 15 at the most.

Figure 10.7 shows how this works for a few input pulses and we urge you to extend the figure to self-test your understanding of the operation of the circuit. Notice that we have assumed that the flip-flop outputs are all at logic 0 before the first upward input pulse transition occurs. We have thus assumed that the counter has been *cleared* before starting operations. How is this managed? Figure 10.5(b) shows that the flip-flops can be cleared initially by bringing R high (assuming that S is low throughout).

Figure 10.7   Sequential action of the counter shown in Figure 10.6.

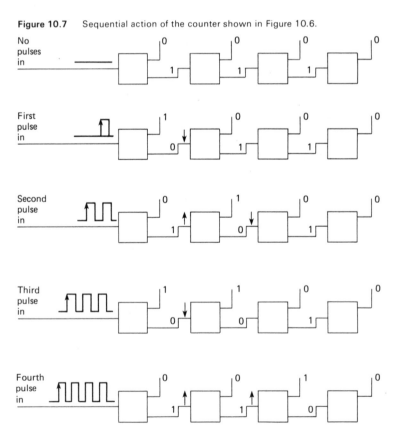

Division-by-*n*

In the circuit in Figure 10.6 output A of FF1 goes 0 → 1 for every *second* 0 → 1 transition of the input signal Clock. Thus FF1 has an output frequency which is one half the frequency of the input signal. We say it has accomplished division by 2. Outputs B, C, and D of the other flip-flops divide by 4, 8, and 16, respectively.

A more general problem is to convert a chain of flip-flops into a divide-by-*n* counter, where *n* has some value that is not a power of 2. A particularly useful instance arises when $n = 10$, because in this instance a modulo-16 counter such as that shown in the figure is converted into a modulo-10 counter, or decade, counter. We notice that as the A, B, C, and D outputs progress during the counting through the decimal numbers 7, 8, 9, and 10, the corresponding binary outputs 0111, 1000, 1001, and 1010 have the special feature that when the count changes from 1001 to 1010, outputs B and D become high simultaneously for the first time. Thus the appearance of $B = D = 1$ is a signal that the count has arrived at 10 (decimal). In Figure 10.8, we show how the addition of one gate to the set of flip-flops will provide a one-bit signal which can be applied to the R inputs of the flip-flops, with the result that the system will be reset to 0000, provided S is held low throughout. The count of 1010 (decimal 10) will never occur. Instead, the decade will begin over again. After you understand this illustration, you should tackle a similar problem, such as converting the modulo-16 counter into a divide-by-7 counter.

You should know that the highly popular 7490 BCD decade counter has as easy-to-use features the ability to divide by 2 and to divide by 5. (See the TTL BCD Counter of Chapter 8.)

An example in which frequency division (divide-by-*n*) is important is given by a battery-operated clock. A high-frequency oscillator controlled by a crystal is used and the output is divided down. The crystal is often the 3.579545 megahertz crystal manufactured for use in color television sets. Division by five decade counters and two divisions by 6 gives a frequency of 0.994 hertz. This is very nearly 1 Hertz.

Figure 10.8   Conversion of a modulo-16 counter into a decade counter.

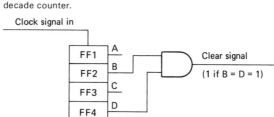

The Master/Slave Flip-Flop

We saw in Chapter 5 that some time is required before an output pin of an IC can respond to changes in the states of the input pins. In a complex system in which gates and flip-flops and other devices are interconnected the various propagation times may result in confusion. An output pin of a particular gate will be definite *if* all the input signals to the gate arrive from the feeding devices simultaneously. However, delays of various lengths of time in the arrivals of the input signals may cause unwanted changes in the output. If this confusion occurs in a complex system, it is called *signal runaround* or *signal race.*

If several flip-flops were interconnected along with logic gates and other devices, there could be serious signal race in the circuit. The solution to the problem in the case of flip-flops has been the introduction of *master/slave flip-flops.*

In a single master/slave flip-flop unit there are two flip-flops which are cascaded in such a way that when the clock goes low (say) data is fed into the master part while at that time the slave part does not accept data from the master part, and when the clock goes high, the master part does not accept data at its inputs, but the slave part is coupled to the master part and the data which the master part received in the previous stage is transferred to the slave part. Step-by-step transferral of data in response to the clock pulses contributes to reliability of operation.

This idea is illustrated schematically in Figure 10.9. In part (a) the master of an RS master/slave combination is coupled to an external signal source but the slave is decoupled from the master. In part (b) the master is decoupled from the source of signals while the slave takes in the data received by the master in the previous step, and adjusts its outputs accordingly.

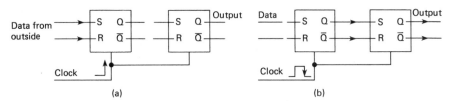

Figure 10.9   The action in a master/slave JK flip-flop.

Ripple and Synchronous Counters

We consider again the chain of flip-flops shown in Figure 10.6 and examined somewhat more fully in Figure 10.7. A change in the output of the first flip-flop obviously occurs only *after* a suitable change in the input of that flip-flop occurs. The second flip-flop cannot determine what its output state must be until the first flip-flop has reacted and signalled to it. Similarly, each flip-flop further down the chain must wait until the preceding flip-flop has reacted. This *ripple* down the chain results in the name *ripple counter*. The 7490 (see the TTL BCD Counter section of Chapter 8) is a much used ripple counter.

If each flip-flop in a chain is a master/slave flip-flop and all are driven by the same clock signal, all of the flip-flops adjust their states in unison. The result is a clocked system which is referred to a *synchronous* counter. The 74193 (TTL BCD Counter section of Chapter 8) is a synchronous counter.

In low-frequency operations such as those which a beginning student is likely to experiment with no problems arise which depend on the kind of counter used. The 7490 and 74193 are equally easy to use. However, in high-speed circuits ripple counters give rise to a problem. Because of timing considerations a circuit using ripple counters may show unwanted state changes which can be attributed to voltage spikes or glitches. This is a well-known problem, but one which goes beyond the bounds of this book.

Suppose that two or more counters are to be cascaded so that together they can count beyond the limit of 9 for a decade counter or the limit of 15 for a modulo-16 counter. This is easily done with 74193 counters which have carry outputs (and borrow outputs for use in down-counting). The 74193 is an example of a *unit-cascadable counter*. The 7490 on the other hand is not of this kind, although two 7490s can be cascaded if some logic gates are used to interface them. These particular devices illustrate another distinction between the two kinds of counter.

It is also possible to cascade synchronous counters so that each is truly synchronous, but so that signals must ripple from each to the following one. This kind of circuit makes a *hybrid counter*.

Some Examples of TTL Flip-Flops

The 7473 is a dual JK device. The pin diagram is shown in Figure 10.10. Each of the two flip-flops in the package has the expected clock, J, and K inputs, and the expected Q and $\overline{Q}$ inputs, but not the expected S and R inputs. One use of the S and R inputs is to clear output Q, as shown in lines 6 and 7 of part b in Figure 10.5 on page 103. The 7473 has instead a single clear input pin for each of the flip-flops. This must be brought low to clear it.

The 7473 is a *level-triggered* TTL device. The states of the outputs Q and $\overline{Q}$ change when the clock input achieves its low or high level.

Figure 10.11 shows another D flip-flop, the 7474. This is a dual edge-triggered flip-flop with preset and clear. It is rising-edge triggered, although the name does not indicate that fact. Figure 10.3(b) leads us to expect clock, D, S, and R inputs, and Q and $\overline{Q}$ outputs. These expectations are fulfilled, except

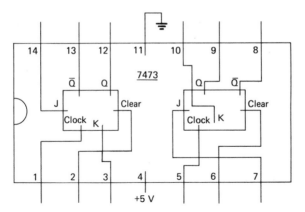

Figure 10.10   Pin diagram of the 7473 dual JK flip-flop.

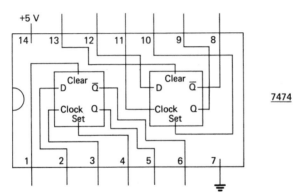

Figure 10.11   Pin diagram of the 7474 dual D flip-flop.

that the clear and set functions which would be performed by S and R inputs are taken care of by a clear input and a set input. For normal operation, each of these should be in the high state.

Having shown examples of TTL JK's and D's, we turn for completeness to a popular TTL RS flip-flop. Figure 10.12 shows the pin diagram for the 74279 quad $\overline{SR}$ latch. Each package contains four flip-flops, but two are two-input flip-flops and two are three-input flip-flops. Another variation from Figure 10.1(b) is that the inputs are $\overline{S}$ and $\overline{R}$, rather than S and R. Finally, no $\overline{Q}$ output is provided. The behavior of the two- and three-input flip-flops and the significance of having the inputs as $\overline{S}$ and $\overline{R}$ can be succinctly explained by truth tables. These are given as the two parts of Table 10.1.

In the pin diagram of Figure 10.12 the internal logic gates which are used to make the flip-flops are shown. The NAND-gate structure should be compared with the circuit shown in Figure 10.1. As an exercise you should show that the S and R inputs of Figure 10.1(b) become $\overline{S}$ and $\overline{R}$ inputs in Table 10.1.

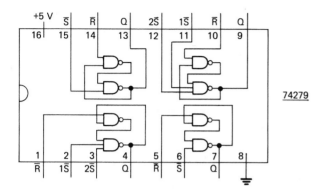

Figure 10.12   Pin diagram of the 74279 quad RS flip-flop.

Table 10.1 Truth tables for the 74279 quad RS flip-flop.

Inputs		Output		Inputs			Output
$\overline{S}$	$\overline{R}$	Q	$1\overline{S}$	$2\overline{S}$	$\overline{R}$		Q
1	1	Latched	1	1	1		Latched
1	0	0	1	1	0		0
0	1	1	0	0 or 1	1		1
0	0	Not allowed	0 or 1	0	1		1
			0	0 or 1	0		Not allowed
			0 or 1	0	0		Not allowed
Two-input flip-flop of 74279			Three-input flip-flop of 74279				

Latching

It is often necessary to hold, or *latch,* some binary information. An example occurs when a counter feeds BCD data to a decoder which drives a display unit such as a seven-segment display. If the BCD data are changing rapidly the display unit segments will go on and off rapidly and the eye will perceive an 8. In order to prevent this it would be nice to be able to command the system to latch a number on the display for some length of time. Also, you may want the states of the output pins on an IC to be latched, aside from the display problem. This is so in memories, for example.

The RS flip-flop can be used as a latch. The fourth line in Figure 10.1(b) can be interpreted in this way: The commands S = 0 and R = 0 tell the RS to hold unchanged at its Q output the last binary digit which was put there. Hence the fourth line in the table could read "S = 0; R = 0; Q latched; $\overline{Q}$ latched."

109

Let us envision a system which is made up of an RS flip-flop and whatever gates are needed to obey the state table of Figure 10.13(a). Figure 10.13(b) shows in schematic form the nature of the system. The double-line rectangle encloses the system, inside which there will be an RS flip-flop and other devices.

This system has two inputs and two outputs but we will use output Q only. There is assumed to be a stream of incoming 0's and 1's at the Data in input. These might come from one of the outputs of a counter, for example. The Control input is to be such that if it is brought low the output Q will follow the Data in signal, and if it is brought high, output Q will latch whatever Data in showed at that moment.

A way to accomplish this is shown in Figure 10.14(a). In the accompanying table [Figure 10.14(b)] the information in Figure 10.13(a) is repeated, along with two columns which show how the S and R inputs to the flip-flop react to the Control and Data in inputs.

By bringing Control low, the output will follow Data in. By bringing Control high, the output that appears then will be latched. Notice that the S and R inputs to the RS are never brought into the states $S = 1$ and $R = 1$.

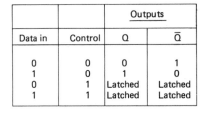

Data in	Control	Outputs	
		Q	$\bar{Q}$
0	0	0	1
1	0	1	0
0	1	Latched	Latched
1	1	Latched	Latched

(a)

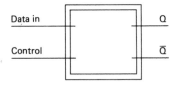

(b)

Figure 10.13   (a) Truth table for a latch circuit based on an RS flip-flop; and (b) schematic showing inputs and outputs of the circuit to satisfy (a).

Figure 10.14   (a) Some inner details of the circuit shown in Figure 10-13 (b); and (b) a truth table for (a).

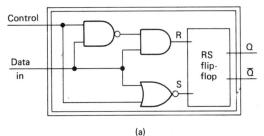

(a)

Control	Data in	S	R	Q
1	1	0	0	Latched
1	0	0	0	Latched
0	1	0	1	0
0	0	1	0	1

(b)

A D Flip-Flop in Use

In our earlier discussion of the D flip-flop we used the phrase "sample-and-hold." As an example of a use for a D we will show how this function might be put to use. Suppose that there is a series of 1's and 0's on some wire. These data might, for example, be generated when automobiles pass a check point. Suppose that what is desired is to sample this stream of 1's and 0's periodically. The purpose might be to measure for a stop light timing change or it might be to collect sampled data for a statistical analysis. We will not attempt to deal with why the sampled data are processed, but only with the sampling itself.

The scheme is indicated in Figure 10.15. On the line to the D input of the flip-flop are somewhat random pulses. The Clock generates a sequence of pulses which are regularly spaced in time. (In a variation of this scheme, the clock pulses could be generated in some other time-dependent fashion.) With S and R set at logic low (or with *set* and *clear* not activated) output Q will react as shown by producing the states shown at the Q output pin.

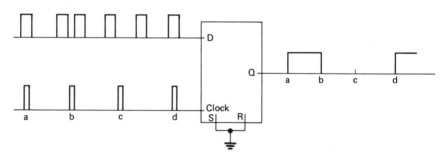

Figure 10.15   Use of a D flip-flop to sample the D-input data on command.

The Internal Logic of a Counter

In this book we rarely discuss the transistors and other circuitry which are actually present in a chip, and then only in sketchy fashion. (The discussions in Chapter 15 are somewhat more detailed.) Similarly, we have not described the internal logic or *functional blocks* which are in an IC except in a few special instances. However, we think it instructive at this point for you to examine the internal logical structure of a counter. Figure 10.16[1] is provided for this reason.

[1]From "TTL Catalog Supplement From Texas Instruments," Texas Instruments Corporation, page S8-6. 1970. Courtesy of Texas Instruments, Incorporated.

functional block diagram

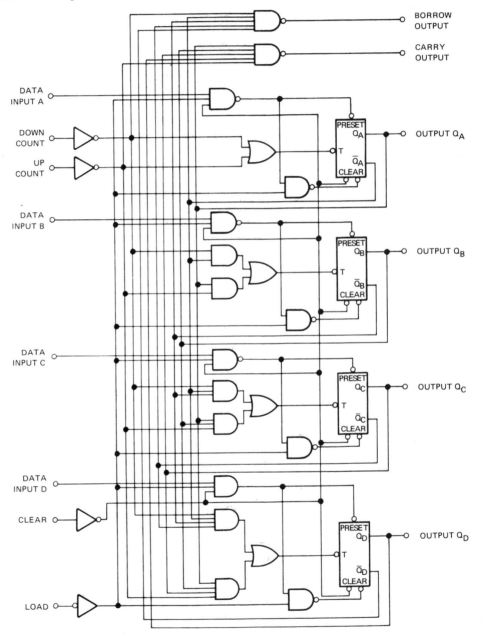

Figure 10.16 Complete internal logic of a synchronous four-bit up/down counter.

The counter in the figure is the 74193 (or 54193). Reference to Figure 8.3 and comparison of the pins shown there with the named functions at the left and at the right of Figure 10.16 may be helpful. The casual reader may be satisfied to observe that the 74193 consists of some logic gates and four flip-flops. (The latter is in agreement with the fact that this is a four-bit BCD counter.) The more dedicated reader may wish to trace out some of the inter-connections to see in detail how particular pin functions work.

The T inputs are so labeled because they are used to make their flip-flops toggle. The only other unfamiliar element in the figure is the appearance at the lower left corner of the symbol for a noninverting buffer with a circle at the input side of the symbol. In this circuit inputs are made active by bringing them high in all cases except in the case of *load.* Load is made active by bringing it low and that is what the circle at the input to the buffer signifies.

EXPERIMENTS

1. A basic RS flip-flop experiment

For a straightforward but useful experiment, try the circuit shown in Figure 10.1(a). This is not entirely trivial, because of the peculiarities in the truth table shown in Figure 10.1(b), and discussed in the accompanying text.

We also suggest designing and building an RS flip-flop using NAND gates rather than the NORs used in Figure 10.1(a). Does your circuit also show peculiarities such as states to be avoided?

2. Latching with an RS flip-flop

We suggest building the circuit shown schematically in Figure 10.14(a). To say that the RS flip-flop can be used as a latch (as is said in the Latching section) seems simple, but to use an RS with only a Control and a Data in input is significantly different.

3. Some basic JK flip-flop experiments

As a first experiment wire up a single JK flip-flop, using a debounced switch as a way to feed in manually controlled input pulses, and using an LED or other simple logic indicator at the output. Among the suitable TTL devices are the 7473, the 7476, and the 74107. The pin diagram for the 7473 is given in Figure 10.10. The pin diagrams for the others are given in Figures 10.17 and 10.18.

You should not only observe the flip-flop action, but ask whether the device shows by its behavior that it is a leading edge or a falling edge or other kind of flip-flop. Trying the J, K, and clear inputs are essential. Instead of using a debounced switch at the input, try just connecting one end of a wire lead to the input pin of the flip-flop, and alternately touching the other end of the wire

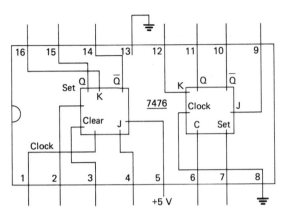

Figure 10.17 Pin diagram of the 7476 dual JK flip-flop.

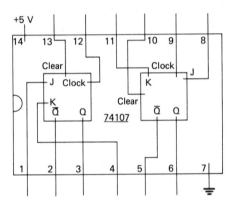

Figure 10.18 Pin diagram of the 74107 dual
JK flip-flop.

lead to the +5-volt bus and breaking that connection. Is the flip-flop sensitive
to switch bounce?

As a further step, make a divide-by-4 counter, using *two* JK flip-flops. The
simplest arrangement is to drive the counter with a debounced switch, and to
watch the output to verify that the circuit is doing what it is intended to do. To
go further, drive the counter with a low-rate clock (such as that described in
Experiment 1 in Chapter 6), with logic state indicators to show the states of the
clock (input) line and the states of the two outputs of the system, and then
while the action is going on, try grounding the clear input, the J input, the K
input, or combinations of those. To witness an essential point about the

memory capability of the counter, disconnect the clock while the system is working, and see whether or not latching takes place. Whether it does or does not, what is the reason in either case? Also: Does the count resume properly when the clock is reconnected?

4. A four-bit BCD counter, using four flip-flop stages

This is an experiment with JK flip-flops. It is recommended, whether any other flip-flop experiments are undertaken or not. The idea is to connect up four JKs to make a modulo-16 counter. This requires two dual JK TTL packages such as the 74107, 7473, or 7476. Pin diagrams for these are given in this chapter. The concepts needed to design the circuit are given in Important Uses for JK Flip-flop: A Binary Counter.

In order for the circuit to work properly, the J and K input pins of the flip-flops must all be at logic high. Suppose these pins are left dangling—not directly connected to the +5-volt bus. While this would seem to be in accord with the general principle that a TTL device input pin that dangles acts as though it were at logic high, you may find that an exception occurs in this circuit. If the circuit does not count properly, no other mistakes have been made, and if the J and K pins have been left unconnected, you should use a logic state indicator to determine what the logic states of those pins really are. Tying the J and K pins to the 5-volt bus should cure the trouble.

5. Latching with a D flip-flop

Figure 10.19 shows the essential concept of a circuit for studying the latching of a counter. The counter can be a 7490 or an other BCD counter. The latching is provided by a 7474 dual flip-flop. The 7475 quad latch might be tried.

Your purpose in the experiment is to verify that the latching action is there. You should be able to command the display to hold the count it was showing when the command was given, and later to command resumption of display.

Figure 10.19   Latching with a D flip-flop.

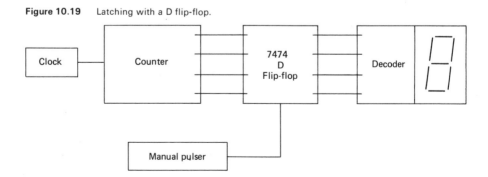

6. The sample-and-hold feature of a D flip-flop

The basic idea of the circuit is shown in Figure 10.20. The circuit is simple because the function of the D flip-flop is itself simple—but sometimes very useful!

The purpose is to observe the ability of the D to look at the signal at the data input pin on command, to set its Q output correspondingly, and to hold its output until commanded to look again at the data input. In order to make it easy to follow the action, the state of the data input pin and the commands at the clock input pin are both manually controlled. A good idea is to predict on paper what the output Q should be after some sequence of operations of the two manual switches. Then try out the whole scheme with the actual circuit and compare what really happens with the theoretical predictions.

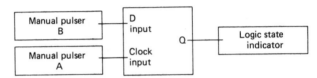

Figure 10.20   The nature of the circuit for Experiment 6.

11

Multiplexing
and Associated Ideas

Multiplexing and Demultiplexing in General

In *multiplexing* any one of several data sources is connected at any one time to a single output line. The situation in Figure 11-1 illustrates the idea. A space vehicle has aboard several sets of scientific instrumentation. Let us assume that one measures the ambient magnetic field magnitude, another the cosmic ray intensity, and that the others (of which we assume there are only two for simplicity) measure other physical quantities. All the data are ultimately in the form of trains of binary digits as they come from the instrument packages.

The space vehicle carries but one radio transmitter, which is used to send the data back to earth. A multiplexer (or *MUX*) is called for. Its function is to connect the data input lines one at a time to the data output line which leads to the single radio transmitter. The essential idea is illustrated in Figure 11.2.

In one version of the system, some electronics might throw the switch in Figure 11.2 from a to b to c to d and then repeat the cycle over and over at regularly timed intervals. This could be called "consecutive sequencing." In another version the sequencing might contact the input lines cyclically but in some other order. In still another version the input lines might be switched to the output line only on receipt of suitable commands from a transmitting station on the Earth. As the data arrive consecutively in time at the Earth the reverse problem arises. Let us suppose that the magnetic field data are to be shunted off to one receiving point (such as a special tape recorder), the cosmic ray data to another receiving point, and so on. A *demultiplexer* can solve this derouting problem. Figure 11.2 will show the idea if the words "Input" and "Output" are read as "Output" and "Input."

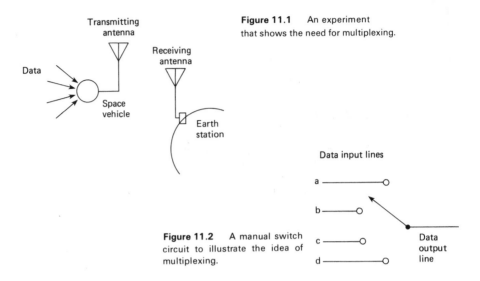

Transmitting
antenna

Receiving
antenna

Data

Space
vehicle

Earth
station

Figure 11.1   An experiment
that shows the need for multiplexing.

Data input lines

a

b

c

d

Data
output
line

Figure 11.2   A manual switch
circuit to illustrate the idea of
multiplexing.

Other ancillary digital electronic operations would be called for in practice, such as synchronization to get the received signals on an input line at the Earth station at just the right moment and to get them terminated at just the right moment.

TTL Multiplexers and Demultiplexers

Figure 11.3 shows the pin diagram for the 74150 One-of-Sixteen Data Selector. *Data selector* is an alternate name for a multiplexer. Table 11.1 shows the state table for the device.

Figure 11.3   Pin diagram for the 74150 1-of-16 data selector, or multiplexer.

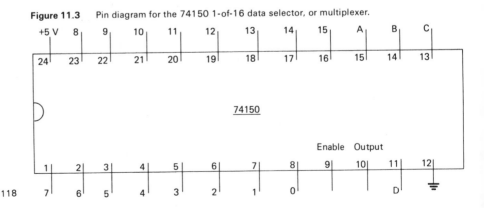

Table 11.1 State table for the 74150 multiplexer.

Address Input				Input Connected	Pin Number of
D	C	B	A	to Output	Input
0	0	0	0	0	8
0	0	0	1	1	7
0	0	1	0	2	6
0	0	1	1	3	5
0	1	0	0	4	4
0	1	0	1	5	3
0	1	1	0	6	2
0	1	1	1	7	1
1	0	0	0	8	23
1	0	0	1	9	22
1	0	1	0	10	21
1	0	1	1	11	20
1	1	0	0	12	19
1	1	0	1	13	18
1	1	1	0	14	17
1	1	1	1	15	16

The output line is connected to that one of the input lines which is specified by the address inputs. For normal operation *Enable* (pin 9) must be low. If Enable is made high, the output goes high. When an input line is connected to the output line, the logic level of the output line is the *couplement* of the logic level of the input line at any time. (An inversion takes place inside the chip.)

The 74154 is a one-of-sixteen data distributor, or demultiplexer. This TTL device connects the input to that one of the sixteen outputs which is selected by the Output select (or Address) inputs. The pin diagram is given in Figure 11.4. The 74154 follows the state of the input pin without inversion at the selected output pin. In normal operation Enable is held low.

Because multiplexers and demultiplexers are more specialized devices than are the devices we have been discussing until now, we limit ourselves to describing some more without giving pin diagrams.

The 74151 is a one-of-eight data selector. This 16-pin chip is basically a trimmed-down version of the one-of-sixteen data selector. With pins to spare, a $\overline{Q}$ as well as a Q output is provided.

The 74153 is a dual one-of-four data selector. This chip contains two independent one-of-four data selectors. Each selector has its own enable, output, and input pins. However, the two selectors share in common the

Address pin.

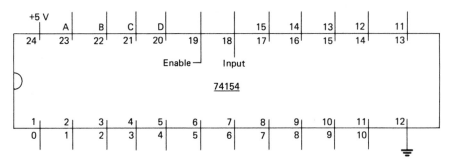

Figure 11.4   Pin diagram for the 75154 1-of-16 data distributor, or demultiplexer.

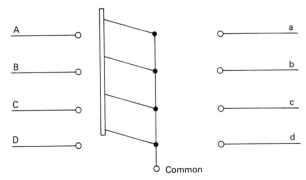

Figure 11.5   A manual switch circuit to illustrate the idea of the quad 1-of-2 data selector.

The 74155 is a dual one-of-four data distributor. This is basically a scaled-down version of the one-of-sixteen distributor. The units share the output select pins. A quirk of the 74155 is that one of the units inverts the data while the other does not.

The 74157 is a quad one-of-two data selector. What the chip does is represented simply in Figure 11.5. It acts like a four-pole, double-throw switch. By throwing the switch arm to the left all of A, B, C, and D are connected to Common, and by throwing the arm to the right all of a, b, c, and d are connected to Common.

An Example of Sequencing

The circuit of Figure 11.6 shows how a clock and a 7490 counter can be used to sequence a 74154 demultiplexer. Since the 7490 is a decade counter, this

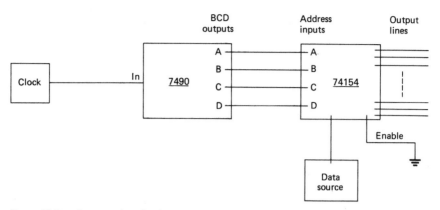

Figure 11.6   A sequencing circuit.

circuit can address only 10 of the 16 output lines of the 74154. If a 7493 or 74193 counter were used, all 16 of the output lines could be addressed.

Multiplexing with Three-State Buffers

Consider a computer in which there are numerous memory devices and a central processing unit (CPU). If the memory outputs were connected by wires directly to the central processing unit with which they must communicate, the number of interconnecting wires would be very large. There would also be a problem in that only one of the memories can be connected to the CPU at any one time in order to avoid confusion. The solution calls for multiplexing. This can reduce the number of connecting wires drastically and it can also solve the problem of conflict.

The three-state buffer provides a way to accomplish the purpose. The multiplexing is done in this case without using ICs which are *explicitly* multiplexer devices of the kind we have been discussing.

The circuit[1] shown in Figure 11.7 shows the underlying idea. The circuit multiplexes two counters to a decoder which drives a seven-segment display unit. The purpose in using two different counters, one a decade counter and the other a full binary counter, is that you can tell by watching the display which of the counters is multiplexed in.

[1] From *The Bugbook II,* by D.G. Larsen, P.R. Rony, and R.A. Braden, pages 7–8E. E. & L Instruments, Inc., Derby CT. Copyright © 1974 by Peter R. Rony. Used with permission.

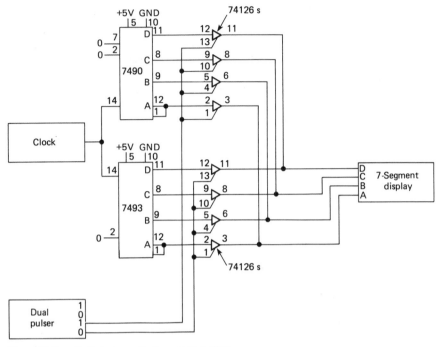

Figure 11.7   Multiplexing with three-state buffers.

Multiplexing a Set of Seven-Segment Displays

In calculators, digital clocks and watches, and many other devices there is usually a set of seven-segment display units. For example, there may be a set of 12 display units in a calculator. To use one decoder with each of the 12 units would result in an unnecessarily elaborate circuit. Great simplification can be achieved through multiplexing. The basic idea is illustrated in Figure 11.8[2]

The SP4T[3] Selector and the 4P4T[3] Selector are multiplexers, but are represented only schematically. The Scan Oscillator generates signals which cause the SP4T selector to connect the anodes of the display units sequentially to the +5-volt supply terminal, and simultaneously to connect the ABCD outputs of the appropriate counter to the decoder/driver. Thus one and only one of the display units at a time is powered and gets data from the decoder/driver. If the scan oscillator frequency is greater than about 100 hertz, each display unit appears to be continuously lit.

[2]From *TTL Cookbook*, by Don Lancaster, pages 295–96. Copyright © 1974 by Howard W. Sams & Co., Inc. Indianapolis, IN. Used with permission.

[3]"SP4T" means "single-pole, four-throw," and "4P4T" means "four-pole, four-throw." These phrases succinctly describe the natures of the switches they refer to.

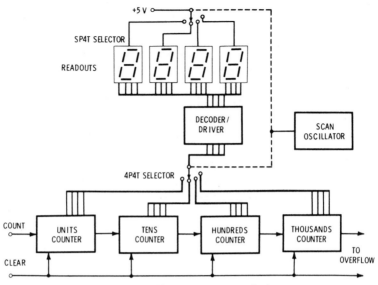

Figure 11.8   Multiplexing a set of four seven-segment displays.

A Multiplexer Used to Generate an Arbitrary Truth Table

A useful trick in designing a digital circuit can be to use a device for a purpose other than its usual one. This can be done with a multiplexer, and we illustrate the basic idea by dealing with a particular problem.

Suppose that in some application the truth table shown in Figure 11.9(a) is needed. This could be accomplished by interconnecting logic gates but there is a considerable savings in complexity (and cost) if a multiplexer is used instead.

Figure 11.9   (a) An arbitrary truth table; (b) a circuit to perform according to (a).

Inputs				Output
d	c	b	a	Q
0	0	0	0	0
0	0	0	1	1
0	0	1	0	1
1	0	0	1	1
1	1	1	0	0
1	1	1	1	1

(a)

(b)

123

How a 16-to-1 data selector can be used in this case is shown in Figure 11.9(b). Input pins 3–8, 10–13, and 16–18 are not used but this apparent wastefulness is justified by the simplicity of the scheme. Of course, there is a limitation to the applicability of this kind of circuit in that the truth table can have only one output column.

Devices other than multiplexers can be used to produce arbitrary truth tables. Decoders have been in that way and since they have more outputs than the multiplexer in Figure 11.9, truth tables which are more complicated can be achieved. As an exercise, you might ask yourself how you can use data selector logic to make a full-adder circuit.

EXPERIMENTS

1. An experiment with sequencing
A circuit that sequences a multiplexer is shown in Figure 11.6. You need logic state indicators for detecting the sequencing process. (These are not shown in the figure.) You may prefer a manually operated pulser instead of a free-running clock at the input to the counter.

2. Multiplexing and demultiplexing
Figure 11.10 shows a simple circuit in which both multiplexing and demulti-plexing take place. This is, in fact, a small-scale model of a telephone switching system. The incoming signal from the clock is routed through a 74155 and a 74153. The clock pulses are received at the output of the system and indicated by the LED there only if the multiplexer and demultiplexer are at the same address. The purpose of the intermediate LEDs is to verify that the signal is being routed as expected. The pin diagrams for the 74153 and 74155 are shown in Figure 11.11.

Figure 11.10   Multiplexing-demultiplexing circuit for Experiment 2.

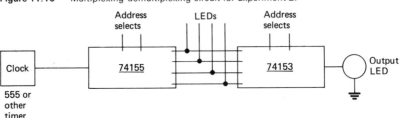

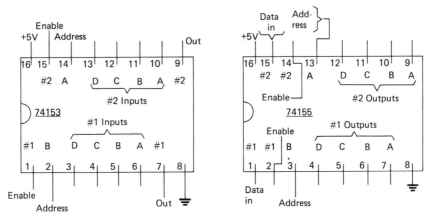

Figure 11.11 Pin diagrams for the 74153 and 74155.

3. Multiplexing with the three-state buffer

Figure 11.7 of the Multiplexing with Time-State Buffers section of this chapter shows a circuit that provides an interesting experiment. This circuit provides for selecting on command which of two different counters is to be connected to a decimal-reading seven-segment display unit. It is therefore a multiplexing circuit. However, instead of using a TTL multiplexing device, the circuit uses three-state buffers. The purpose in using a 7490 counter and a 74193 counter is that as you watch the seven-segment display, you can tell whether the decimal 7490 is driving the display, or the modulo-16 74193, and so you can check on the correct operation of the multiplexing.

4. Use of a multiplexer to generate an arbitrary truth table

The table of Figure 11.9(a) shows an example of a four-input, one-output truth table which might be needed in some special application. You might of course try to design a circuit using the basic logic gates to make a system that will obey that truth table. This can indeed be done, but in the rich world of IC devices there may be a trick available that simplifies the job. As Figure 11.9(b) shows, a multiplexer can be used, with an assured gain in simplicity over a circuit that uses only logic gates. We suggest that you invent a similar but different four-input, one-output truth table, and then design and build a circuit based on a multiplexer that will operate according to your truth table.

12

Memories

Extensive electronic memories date back only to about 1950. A notable early use was in making calculations needed for the construction of the first hydrogen bomb.[1] But today even a casual look at the world about us should convince you that memory devices are used by virtually everyone. The uses span such areas as electronic calculators, banking, supermarket checkouts, and numerous control applications.

Today the trend is toward memories with greater storage capacity, higher operating speed, smaller size, and lower cost. Advances are being made rapidly and great improvements can be expected. Already a few memory chips of almost negligible size can replace what would have required a roomful of electronic components not many years ago.

Memory devices exist in considerable variety. There are magnetic tapes, magnetic cards to plug into calculators, the discs used with microcomputers and large-scale computers, and the up-and-coming bubble memories, to mention only a few. Our concern will be with integrated circuit memories, and with TTL memories in particular. In Chapter 15 we will discuss the nature of CMOS, and you will be able to carry over the ideas in this present chapter to understanding CMOS memories.

Classification of Memory Devices

Before reading about specific memories, you should understand how memories are put into classes and you should have in mind the terminology which has grown up.

[1]See *Brighter Than a Thousand Suns*, Robert Jungk, Harcourt Brace Jovanovich, 1958.

One kind of integrated circuit memory is known as a *ROM,* which means *Read Only Memory.* A ROM has data hard-wired into its internal electronics. The data are permanently fixed. You cannot "write" into a ROM, which is to say that you cannot in any way alter the data stored. A ROM could be used in a calculator to store commonly used numbers, such as the values of π and e (the base of the system of natural logarithms) so that you can call forth the number you need at the push of a button, for example. The data stored in a ROM can be program instructions as well as numerical data. Thus a ROM might be used to hold the instructions which are used to control the various modes of operation of a microwave oven.

There are three ways in which it is determined what information will be stored in a ROM. For one, manufacturers of ROM will accept from purchasers the specification of what the contents should be and then turn out the ROMs accordingly. For example, a firm that intends to produce microwave ovens could tell the ROM manufacturer what the contents of the ROMs they need must be. As you would expect, this is economically feasible only for large-quantity ROM users.

Second, some particular ROMs can be expected by the ROM manufacturer to be of widespread utility and the manufacturer may put those specifically programmed ROMs on the general market. An example would be a character generator, the purpose of which is to interpose itself between the keyboard of a microcomputer, and to convert the fact that the user presses a key into corresponding information in some binary code which is suitable for entry into the microcomputer.

The third way in which a ROM can have information stored in it is called *field programming.* Users can buy a blank ROM and put into it whatever they want. A ROM which can be programmed in this way is known as a PROM, which means *Programmable Read Only Memory.* How this is done is discussed later in this chapter. A true PROM cannot have its contents changed once it has been programmed but it would obviously be desirable sometimes to be able to "erase" a PROM and to enter new data into it. Indeed, there is a special kind of PROM with which this is possible. It is known as an *EPROM,* which means *Erasable PROM.* Ultraviolet light is used for the erasure. (See the PROMs and EPROMs section of this chapter.)

A more recent development has resulted in EPROMs which can be erased electrically rather than by the use of ultraviolet light. They are called *EAPROMs* which means *Electrically Alterable PROMs.*

We turn next to a kind of memory device which is of a distinctly different kind. While you can read information out of a ROM but not write information into it, you *can* write into this kind of memory. Such a memory device is a *read/write memory.* While a programmed ROM is somewhat like a stone monolith into which information has been chiseled so that you can only read the information, a read/write device is more like a blackboard on which you can write information and easily replace previously written information with new information. A read/write memory is called for when (for example) the inventory data for a business operation are to be stored and updated as need arises. In a computer the program instructions are written into a read/write memory. Once the instructions are in the memory, the computer executes them step-wise. Later an entirely new program can be entered, replacing the previous one in the memory.

Another distinct way in which memories are classified is according to whether they are *sequentially addressable* or *randomly addressable*. In order to understand the distinction you should begin by thinking of a memory as having *locations* (or "bins" or "cells") in which it holds data. Thus a memory might have locations numbered 1, 2, 3, and so on, in which there are the binary numbers 00111011, 11011100, 11000001, and so on. A *sequential memory* is one in which the numbers can be referred to by going to the locations only strictly in sequence. As an example, consider data stored item after item on magnetic tape, and a tape reader which can only start at the beginning of the tape, read through to the end, and then begin again. This would be accessing the data sequentially. On the other hand if the data were on a disc and the disc reader could on command jump from wherever it might be along the disc to any other point along the disc, the access to the data would be *randomly addressed*. This would be a case of a *random access memory, or RAM*. Integrated circuit read/write memories are always RAMs.

A critical advantage of a RAM over sequential memory is that in the case of the RAM the *access time* is independent of the location of the data sought. In short, RAMs are inherently faster than sequential memories. They may be *much* faster, in fact.

Here we insert an explanation of the distinction between a *static* and a *dynamic* RAM, but only somewhat parenthetically since you are rather unlikely to choose to work with a dynamic RAM although you should be aware of the meanings of the terms. In a dynamic RAM, the internal circuitry incorporates capacitors. When data is stored in the RAM certain capacitors are charged up. As is the way of capacitors, they tend to discharge exponentially over time. To prevent loss of the stored data, the RAM has to have associated with it some circuitry which periodically *refreshes* the stored data. This refreshing of the data is not necessary with a static RAM.

You must also distinguish between *volatile* and *nonvolatile* memories. *Volatile* memories hold their stored contents only as long as power to the memory unit is supplied continuously. TTL RAM memories are volatile. If power to the chips is lost accidentally or on purpose, the information is lost. On the contrary, a *nonvolatile* memory retains its data without regard for most external conditions. For example, a magnetic tape on which spots of magnetized or unmagnetized material represent data holds its stored data as long as its physical integrity is not destroyed and as long as its magnetic environment is not altered drastically.

An Example of a RAM:
The 7489 64-Bit Memory

The 7489 is a small but representative example of a TTL RAM. Figure 12.1 shows the pin diagram. Figure 12.2 shows the 7489 in a schematic form which should help in understanding the following discussion of the device.

The 7489 holds its 64 bits organized as 16 words,[2] each with a length of four bits. Each of the 16 words is held in one of 16 locations, or addresses. These addresses are numbered 0000 through 1111. The four bits which can be held at each address are called (as usual) A, B, C, and D. In order to write a word into a particular address the word must appear at pins 4, 6, 10, and 12, the device must be supplied the address at pins 1, 13, 14, and 15, and the

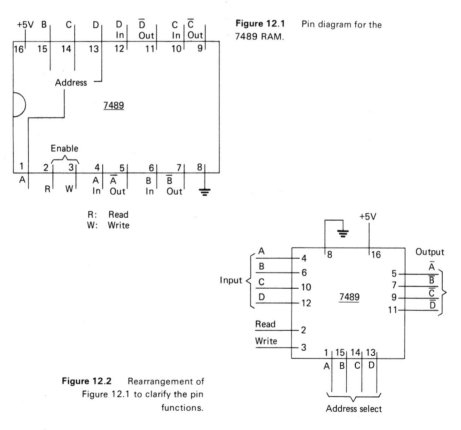

Figure 12.1   Pin diagram for the 7489 RAM.

R: Read
W: Write

Figure 12.2   Rearrangement of Figure 12.1 to clarify the pin functions.

[2]A bit is one binary digit. It is a 0 or a 1. What a *word* is varies with the context. In microcomputers it may be a set of eight bits or a set of 16 bits. In some computers it is a set of 32 bits. In the 7489 it is a set of four bits. Sometimes a set of eight bits is called a *byte,* and then each four-bit half of a byte is a *nibble.*

129

device must be told that writing (rather than reading) is to be done. If a word is to be read out, the address of the word must appear at the address select pins, and the device must be told that readout is what is wanted. When reading out, the information that was stored is not lost, but in writing into any address the information which is entered replaces or "writes over" whatever was there before.

The write enable pin 3 and the read enable pin 2 are used as follows: If both are brought high, the outputs of the 7489 are decoupled from whatever they are connected to. (This is reminiscent of the three-state buffer.) Otherwise the read enable input is left low. If the write enable pin is also brought low, the data are entered into the addressed cell. If the write enable pin is brought high, readout takes place.

If you examine Table 12.1 while the preceding discussion is read through again, the operations should become clear.

A peculiarity of the 7489 is the appearance not of the word itself at the output pins, but of the complement of the word, as the table shows. Inverters at the outputs would provide direct readout rather than complementary readout. Alternatively, the complement of a word which is to be in storage could be read in initially.

To illustrate how all this works, we describe how a telephone number can be put into a 7489 and then called back later. The circuit used is shown in Figure 12.3. Let us suppose the number is 2526275. In BCD form this is 0010 0101 0010 0110 0010 0111 0101. To write in 0101 (the least significant word) the data inputs are set as shown in the figure. If this word is to be entered in location 0000 as we assume, the address is set as shown. This writes 0101 into location 0000. To enter 0111 the write enable pin is brought high, the input pins are set at 0111 and the address pins are set to 0001 to use the next location in sequence. By bringing the write enable pin low the word is written in. This procedure is followed for the rest of the words (digits in the phone number).

Table 12.1 The logic states of the Memory enable and Read/write enable pins and their consequences at the output pins of the 7489 RAM.

Memory enable (pin 2)	Read/write enable (pin 3)	Operation	Output state
0	0	Write	All outputs at logic 1
0	1	Read	Complement of chosen word
1	0	Hold	All outputs at logic 1
1	1	Hold	All outputs at logic 1

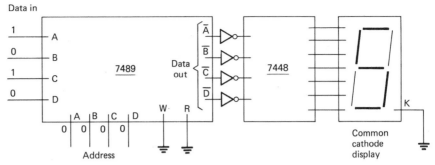

Figure 12.3   Writing 0101 into memory location 0000 of a 7489 RAM.

Once the telephone number is in the memory, it can be called back by sequencing the address pins (probably in the order from 0110 down to 0000 in order to read out 2526275 from the leftmost 2 down to the rightmost 5), holding the write enable pin high and the read pin low.

In this example we have had in mind a labbench set-up for experimental purposes. The procedure described has not made use of the option of bringing W and R both high, an option which may have to be used carefully when some other circuitry is connected to the output pins of the 7489.

An Example of a ROM: The 7488

The stored contents of the 7488 ROM are permanently entered into the device at the factory according to the specifications of the purchaser. It is a 256-bit memory, with the bits arranged as 32 words of 8 bits each. Of the 16 pins, two are needed for the ground and +5-volt connections. Five are needed as address pins, since addressing 32 locations requires the binary numbers 00000 through 11111. Because the data words are 8-bit words there are eight output pins. The remaining pin is the enable pin. Normal operation of the ROM follows if the enable pin is low. If the enable pin is high all of the outputs go high. The pin diagram is shown in Figure 12.4.

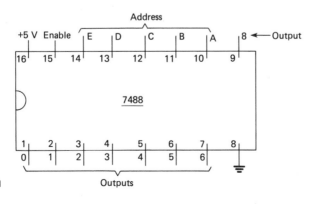

Figure 12.4   Pin diagram for the 7488 ROM.

An Example of a Large RAM

The National Semiconductor DM93415 is a TTL 1024 × 1-bit RAM. The number 1024 is sufficiently close to the number 1000 that such a memory is often referred to as a *1K memory*. The 1024 locations in this RAM can be thought of as forming a 32 × 32 array, as suggested in Figure 12.5. The pin diagram is shown in Figure 12.6.

Each memory location holds only a single bit, hence only one input pin and only one output pin are needed. However, to address a memory location requires that a row in the array be specified by a binary number from 00000 to 11111 and that a column be specified by a binary number of the same length. It follows that two sets of address pins are needed with five pins in each set. With a ground pin and a +5-volt pin, the number of pins so far comes to 14. The remaining two of the 16 pins are labeled $\overline{WE}$, which means *Write Enable*, and $\overline{CS}$, which means *Chip Select*. (The bars in the symbols $\overline{WW}$ and $\overline{CS}$ indicate that those pins are active when they are low.)

The write enable pin is used to tell the device whether writing-in or reading-out of data is to be carried out. The chip select pin is used to specify whether the device is activated or not. Suppose that a number of 93415s were

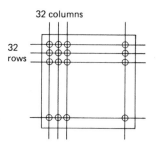

32 columns

32 rows

Figure 12.5 Arrangement of the 1024 memory locations of a DM 93415 in a 32-by-32 array.

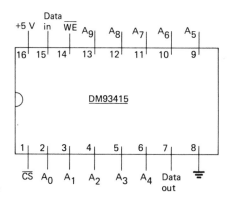

Figure 12.6 Pin diagram for the DM 93415 RAM.

used in a memory system to get a large storage capacity. It would be necessary to specify which of the 93415s is to be used at any one time. $\overline{CS}$ is used for this purpose.

An Example of a Large PROM

The National Semiconductor DM74S573[3] is a TTL PROM. It stores 4096 bits, arranged as 1024 words of four bits each. It is shipped from the factory with 0 bits in all locations. Users must insert 1 bits where they need them. Once programmed it is impossible to change any 1 back to a 0.

Even though this 4K memory is larger with respect to storage capacity than is the 1K 93145, the 74S573 is only an 18-pin device. There are (of course) no input pins. The storage locations must be addressed by 10 pins, there must be four output pins, a ground pin, and a +5-volt pin making a total of 16 pins. The remaining two pins are $\overline{E1}$ and $\overline{E2}$, where E means Enable. For normal operation, each of these must be low and then the output pins present the contents of the selected location. If either of the pins is brought high each of the four output pins goes into a high impedance state. The memory device is decoupled from any loads to which the output pins are connected. (This is the essential feature of the three-state buffer which was discussed in Chapter 4.) The pin diagram for the DM74S573 is shown in Figure 12.7.

Figure 12.7   Pin diagram for the DM 74S573 PROM.

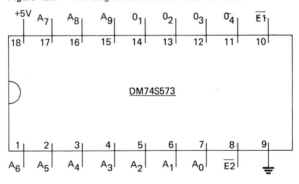

"A" pins are address pins. "O" pins are output pins.

---

[3]A valuable resource book is *Memory Databook,* published by National Semiconductor Corporation, 1977. It describes numerous TTL and CMOS RAMs, ROMs, and other ICs.

PROMs and EPROMs

In this section we intend to give you a general understanding of the methods used in working with PROMs and EPROMs. Readers who want to carry out the actual procedures must have more detailed instructions regarding voltages, currents, pulse times and circuit details.[4] You must also realize that EPROMs are MOS devices (see Chapter 15) rather than TTL devices. In spite of this fact, it is appropriate to discuss them here because they *are* popular memories.

In a PROM as it is supplied by the manufacturer, the critical transistors in the circuitry which are eventually to remember 0's and 1's have fuses in their emitter leads. This is sketched in Figure 12.8. The fuse is usually made of nichrome.

The original PROM may initially have all bits in their 0 states or in their 1 states. We will assume in this discussion that the original PROM has all 0 states.

The basic idea in programming the PROM is to pass a current pulse of suitable shape, amplitude, and duration through the fuse to "blow" it. Then an unblown fuse corresponds to a 0 and a blown fuse corresponds to a 1.

You need the usual DC power supply to power the PROM. You also need a second power supply to furnish the current pulses. To burn a word in, the chip must be properly connected to its power supply and the desired address must be selected by proper 0's and 1's at the address pins. The outputs must be disabled by using the chip select pins. After these steps, a current pulse is fed in. When all the fuses which correspond to the selected address have been processed, the whole operation is repeated for the rest of the addresses.

You can buy commercially-made programming units. These are very convenient but they cost hundreds of dollars, the amount depending on the degree of sophistication. They are practical only where the application justifies the expense.

We turn now to a brief description of the erasure of stored data in an EPROM. An EPROM has on its upper face a quartz window which is typically a rectangle roughly ⅜ inches by ⅝ inches in size. Quartz is used because it

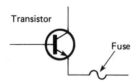

Transistor

Fuse

Figure 12.8　The fuse that is blown when programming a PROM.

[4]An informative article is "PROM Programming," *Radio Electronics,* September 1979, pages 68–71. Another readable article is "How to Program Read-Only Memories" by Robert D. Pascoe, *Popular Electronics,* July 1975, pages 27–30. Still another is "Inexpensive PROM Programmer" by R. E. LeVere, *Radio Electronics,* February 1981, pages 74–77.

passes ultraviolet light, unlike ordinary glass and many other materials which are transparent to visible light but opaque to ultraviolet light. You must have a suitable ultraviolet light source and shine it through the window with sufficient intensity for a suitable exposure time. An efficient way to equip yourself with a suitable lamp is simply to buy one which is sold specifically for this purpose. They can be had for about eight dollars. Ultraviolet light is hazardous, especially to the eyes. Care must be taken when working with it.

EAPROMs

One advantage of an EAPROM over an EPROM is that it can be selectively changed in chosen locations while an EPROM can only be erased entirely and then reprogrammed. Another advantage is that an EAPROM can be manipulated by entirely electrical techniques, thus avoiding the need for ultraviolet light irradiation.

When a previously programmed EAPROM is to have some of its contents changed, you simply enter in the new information where it is wanted, writing over what was there before. The reader who is interested in this must obtain detailed information from the manufacturer, a dealer, or elsewhere.

PROMs are common and readily available on the market and the procedures for working with them are well established. EAPROMs are newer and much less often used.

Use of a Memory to Generate an Arbitrary Truth Table

At the end of the preceding chapter we explained how a multiplexer can be used to replace a circuit made up of logic gates to get a "nonstandard" combinatorial truth table. A memory device can be used in a similar way, and in fact with even greater capacities possible.

To illustrate this idea, we consider how a RAM or a ROM can be used to make a full-adder. Figure 12.9(a) shows the truth table and Figure 12.9(b) shows the circuit. A and B are the bits which are to be added and C is the carry bit from some previous addition which must be added to the sum of A and B.

The memory cells with addresses 0000 through 0111 of a RAM such as the 7489 or of a ROM have previously been loaded with the data 0000, 0010, 0010, 0001, and so on to correspond to the outputs in the table, according to the rule that in each case the A bit represents the sum, the B bit represents the carry, and the C and D bits are not used. The D bit at the address pins is held permanently at 0 and the other input pins are used to address the right memory locations.

Inputs			Outputs	
C	A	B	Sum	Carry
0	0	0	0	0
0	0	1	1	0
0	1	0	1	0
0	1	1	0	1
1	0	0	1	0
1	0	1	0	1
1	1	0	0	1
1	1	1	1	1

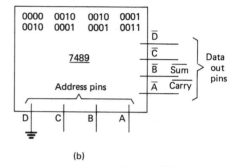

(a)   (b)

Figure 12.9   (a) Truth table for a full-adder; (b) a circuit that obeys the truth table of (a).

This particular example has been chosen for its simplicity, but it is hoped that the possibilities inherent in the idea have been suggested. We suggest that you ask yourself how a ROM could be used as a BCD-to-seven-segment decoder as a test of your understanding of this section.

EXPERIMENTS

1. Data storage with a RAM
This experiment can be important in helping your understanding of memory devices. The circuit is shown in Figure 12.10. This is the circuit given in Figure 12.3 but with a clock and a counter added. The ultimate purpose is to have the clock pulses cause the 7489 to output its stored data in sequential fashion. You must first store the desired information in the 7489.

Figure 12.10   A circuit for writing into and reading out of a RAM (Experiment 1).

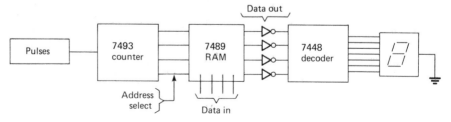

This has proved to be a very instructive experiment for the newcomer to RAMs. Also it is fun to store some special data, such as your Social Security number, and then to watch it come out properly. You might want to go further by carrying out a similar experiment with another RAM, such as that shown in Figure 12.6, but all the ideas involved are illustrated in an experiment that uses the simple 7489 RAM.

2. A full-adder made by programming a RAM

Suppose you need a circuit that obeys some particular truth table and that no IC device of that kind is available. We saw early in this book that the basic logic gates AND, NAND, OR, and so on can be combined in some fashion to make the circuit. In Experiment 4 of Chapter 11 the trick of making a circuit for a special truth table with great simplicity and agreeable elegance consists of using a multiplexer instead of logic gates.

In the present experiment, another trick is used to produce a very simple and appealing circuit, where otherwise the circuit might be complex and awkward. The idea is to use a RAM to make a full-adder circuit, although you might choose to use some truth table other than that for a full-adder. The 7489 RAM is assumed, although any of a number of other RAMs can be used. Writing into and reading from the 7489 is discussed in the second section of this chapter. The use of the 7489 as a full-adder is discussed in the previous section. We hope that you will enjoy building this circuit and watching it work.

The design of this circuit is an instance of what has been called "advanced design techniques" by Don Lancaster in his *TTL Cookbook,* to which we have referred several times. You may wish to consult that book (or others) for further discussion.

13

Shift Registers

A *register* is a digital storage device. An example that we have already met is the four flip-flops in Figure 10.6. This is a register which consists of four individual *stages,* each of which can hold a 0 or a 1.

Like all registers, a *shift register* (SR) also holds 0's and 1's in its stages, but it has the additional property that the contents can be shifted on command. If (for example) a shift register holds 10111 and is a *right shift register,* then on command the leftmost 1 moves over to the second position from the left, the 0 which was originally in the second position from the left moves to the third position from the left, and so on. What happens at each end of the shift register is taken up in the next paragraph. A *left shift register* operates similarly, moving the contents to the left one stage at a time on command. There are also *bidirectional,* or *shift-right-shift-left registers.* The process of commanding the shifting to occur is usually called *clocking*.

In one kind of shift register, as the contents are moved to the output end, the binary digits become available for use in other parts of the circuitry one by one, but each is lost from the register as it is shifted out. The stages of the register which are vacated by the shifting automatically have 0's installed in them. For example, after four shifts out of a four-stage shift register, all of the four-bit information originally held is lost. However, in another type of shift register, each bit is automatically reentered at the other end as it is shifted out at the output end. This is called *recirculation.*

What has just been described is a *serial-out* SR. The digits which were stored originally are read out one at a time at an output pin. All SRs are serial-out but some SRs are, in addition, *parallel-out.* In a parallel-out SR all the data stored in the register at any moment can be read out simultaneously.

The discussion thus far has concentrated itself on the readout of data. The question of entering data into an SR is equally important. In one kind of SR the data to be entered are presented at an input pin one bit at a time in coordination with a clock input pulse. As the clocking proceeds, the successive bits are entered and for each there is a simultaneous shifting of the previously entered bits, each one stage at a time. This kind of SR is called a *serial-in SR*. In contrast there are *parallel-in SRs*. In this type of SR the bits which are to be stored are entered all in one operation.

Combinations of the two kinds of input types and the two kinds of output types lead to *SISO, SIPO, PISO,* and *PIPO* shift registers. For example, with a PISO (parallel-in-serial-out) shift register, the data can be entered simultaneously but the data can be read out only serially.

TTL shift registers of all the four kinds are available with short lengths, such as eight stages. If a very long shift register is needed, practical considerations rule out the availability of all types. For example, a shift register with 2048 stages (a 2K SR) of the PIPO kind would have to have more than 4000 connecting wires for the input pins, output pins, and other function pins. However, a 2K SISO system would need only one input lead, one output lead, and very few other leads. The contrast is striking.

The mention of long shift registers leads to another remark. Short SRs can often be combined, or *cascaded*. For example, to store the BCD representation of a set of four decimal digits, four SRs each with four stages can be cascaded. (One four-stage SR can hold the BCD representation of one decimal digit.) In such cases not only proper interconnections between the registers must be considered, but also the management of the clocking pulses in order to assure read in and read out of data.

Some Uses for Shift Registers

Since a shift register is a memory, it can be used to store information. In large-scale computers and microcomputers, there are always registers used precisely for storage, some of which can be shifted. Another frequent application for SRs is the conversion of data from serial to parallel form, or from parallel to serial form. A SIPO SR can manage the former operation and a PISO SR can manage the other. Sequencing can also be done with shift registers. The clocking of a circulating shift register in which a string of 0's and 1's has been stored will produce a sequence of signals at the output pin of the SR, and the sequence will recur cyclically as long as the string of clock pulses is not interrupted. Among the many other applications which ingenuity can devise is frequency division. For example, if (say) 00001000 is stored initially in a shift register, a 1 will appear at the output for every eighth clock pulse, thus accomplishing a division of the clock frequency by 8. Another application is in the generation of time delays in the transmission of signals. A very common application today is in cathode-ray tube displays in computer terminals. The characters seen in the display may (for example) consist of 5-by-7 matrices. In one scheme, the five dots in a row of a character matrix are loaded in parallel fashion into a shift register, and then the dots are read out onto the computer monitor screen in serial fashion. This "TV typewriter" kind of display is a high-speed operation and TTL shift registers are adapted to it because of their high speed.[1]

A Shift Register Made with Flip-Flops

Figure 13.1 shows a circuit which consists of four D flip-clops. It constitutes a basic SISO shift register.

We suppose that initially each of flip-flop outputs Q_D, Q_C Q_B, and Q_A is at logic low and then we suppose that a series of pulses is applied at the clock inputs. If on the arrival of a clock pulse the serial input is 0, no change will occur in any of the flip-flop output states. If, however, the serial input is 1, output Q_D will become 1. On three following clock pulses this particular 1 will shift to Q_C to Q_B, and then to Q_A. As each shift occurs, the state of Q_D will be determined by the state of the serial input. After the four clock pulses, the shift register is full. At any later time the stored data can be read out serially as Q_A by the application of clock pulses.

Parallel input and output capabilities make shift registers more versatile. We show a way to convert the SISO SR of Figure 13.1 into a SIPO SR. The circuit is shown in Figure 13.2.

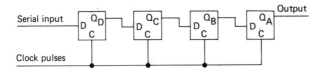

Figure 13.1   A basic SISO shift register.

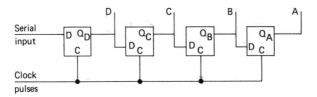

Figure 13.2   A basic SIPO shift register.

[1]This matter and a number of others are discussed readably in *TTL Cookbook* by Don Lancaster, Howard W. Sams and Co., Inc., 1974, page 266 and preceding and following pages. Such topics are dealt with more fully in *TV Typewriter Cookbook* by Don Lancaster, same publisher, 1976.

Some TTL Shift Registers

There are numerous TTL shift registers. We describe briefly one SIPO, one PISO, and one PIPO. The 74164 is a SIPO SR with eight stages. The pin diagram is shown in Figure 13.3. The eight bit output appears at pins 3, 4, 5, 6, 10, 11, 12, and 13. An unexpected feature in the light of what has been said before is that there are *two* input pins, namely pins 1 and 2. If either of these is brought high, the other accepts input data. Thus two sources of input signals can be made use of, with the possibility of complicated mixing. The protection of computer data which is mentioned in the Programmable Sequence Generator section of this chapter is an example of an application.

The 74165 is a PISO SR with eight stages. The pin diagram is shown in Figure 13.4. This device has 12 input pins. Eight of these are for parallel input

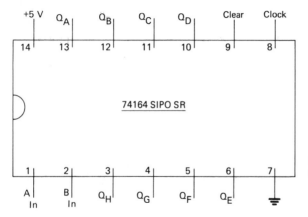

Figure 13.3   Pin diagram of the 74164 SIPO shift register.

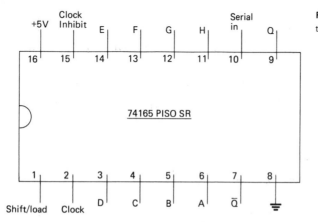

Figure 13.4   Pin diagram of the 74165 PISO shift register.

141

of eight bits and the others are for the clock input, the clock inhibit, the shift/load, and the serial input option. The state table given in Table 13.1 is a succinct way to explain the functioning of these inputs. Wherever the table says "1 or 0" with respect to a spin, the functioning of the device does not depend on the state of the pin referred to.

Table 13.1 State table for the 74165 SIPO shift register.

Shift/load input	Clock inhibit input	Clock input	Serial input	Parallel inputs	Q output	$\overline{Q}$ Output
0	1 or 0	1 or 0	1 or 0	A ... H	A	$\overline{A}$
1	0	0	1 or 0	1 or 0	Last value before input was accepted.	
1	0	0 → 1	1	1 or 0	Value of B before last clock transition. Input H set equal to 1.	
1	0	0 → 1	0	1 or 0	Value of B before last clock transition. Input H set equal to 0.	
1	1	0 → 1	1 or 0	1 or 0	Last value of A before input was accepted.	

The parallel inputs are loaded by bringing the shift/load input low regardless of the states of the clock, clock inhibit, and serial input. Clocking cannot take place unless the clock inhibit is brought low. The serial input is used to change the value of input H only. When the serial input is used, the SR becomes a SISO SR.

We conclude this section with a discussion of the 74195, which is a PIPO shift register with four stages. This chip is quite straightforward to use. To input data, you bring the shift/load input low and the data will be loaded on the next clock transition. If shift/load is made high, the data will be shifted. The pin diagram is shown in Figure 13.5 and the state table in Table 13.2.

In the table, the symbols Q_{AO}, Q_{BO}, Q_{CO}, and Q_{DO} are the values of Q_A, Q_B, Q_C, and Q_D before the input conditions are applied, and Q_{BN}, Q_{CN}, and Q_{DN} are the values of outputs A, B, C before the last clock transition.

To input data, you simply bring the shift/lead input low and the data will be loaded on the next clock transition. To shift the data, the shift/load input must be made high. The clear input will force the four outputs A, B, C, and D low and will set $\overline{A}$ high. You gain versatility through the functioning of the two serial inputs.

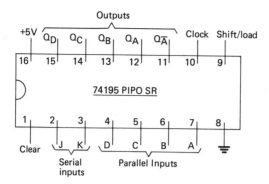

Figure 13.5 Pin diagram of the 74195 PIPO shift register.

Table 13.2 State table for the 74195 PIPO shift register.

Inputs			Outputs			
Clock	J	K	D	C	B	A
$0 \rightarrow 1$	0	1	Q_{DO}	Q_{DO}	Q_{CN}	Q_{BN}
$0 \rightarrow 1$	0	0	0	Q_{DN}	Q_{CN}	Q_{BN}
$0 \rightarrow 1$	1	1	1	Q_{DN}	Q_{CN}	Q_{BN}
0	0 or 1		Q_{DO}	Q_{CO}	Q_{BO}	Q_{AO}

An Example of
Serial-to-Parallel Data Conversion

Figure 13.6 shows a circuit[2] with which data can be read into the system serially and read out in parallel fashion.

The clock signal is applied to a 7490 counter which is wired up to act as a divide-by-8 counter, and simultaneously to the 74164. The data to be loaded are applied serially at the clock rate to the serial input of the 74164. After eight clock pulses the shift register holds the eight binary digits which were input, and the next clock pulse causes the counter to signal the 74374 octal D flip-flop to accept all eight digits simultaneously from the shift register. In the next cycle of eight clock pulses, new data can be entered into the shift register, and on the following clock pulse the new data can be transferred to the 74374.

---

[2]From "More on Shift Registers" by Forrest M. Mims, *Popular Electronics,* November 1980, page 108. Reprinted from *Popular Electronics Magazine.* Copyright © 1980 Ziff-Davis Publishing Company. Used with permission.

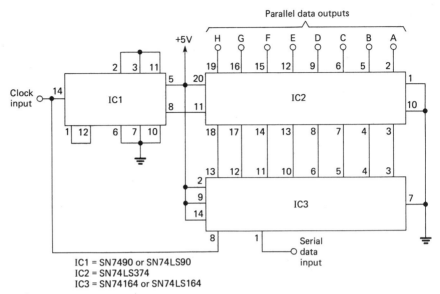

Parallel data outputs

+5V

Clock input

IC1

IC2

IC3

Serial
data
input

IC1 = SN7490 or SN74LS90
IC2 = SN74LS374
IC3 = SN74164 or SN74LS164

Figure 13.6   A circuit for conversion of serial input data into parallel output data.

A Programmable Sequence Generator

The purpose of the circuit[3] shown in Figure 13.7 is to enable you to select any pattern of eight binary digits by means of manual switches, and to have the pattern readout serially in response to clock pulses. This is a PISO operation. The circuit uses two 74194[4] shift registers in series. The clock signals are applied simultaneously to the clock inputs of both shift registers. The result is that the pattern is shifted one position for each clock pulse through the two 74194 shift registers. The serial output of the first 74194 feeds the serial input of the second 74194. The light-emitting diodes are included to make it possible to follow the operation of the circuit by eye, assuming a slow clock rate.

The 0.1-microfarad capacitor which appears between the power supply +5-volt terminal and the circuit ground is to swallow up power supply transients which may occur during operation and which could affect the sequencing.

[3]Ibid., page 109.

[4]The 74194 is a bidirectional shift register.

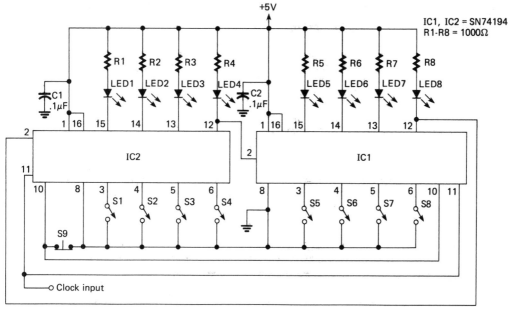

Figure 13.7   A programmable sequence generator.

A Pseudorandom Generator

There are many applications for circuits which put out numbers in a random sequence. These include games which depend on random results, noise generation for audio equipment testing, and cryptography.

A pseudorandom generator differs from a truly random generator in that it produces a highly jumbled sequence of numbers but repeats the sequence over and over cyclically. This is an adequate approach to true randomness for many purposes.

The circuit[5] in Figure 13.8 uses a six-stage shift register consisting of D flip-flops. It produces a sequence of 63 (0 to 62) jumbled numbers. (In general if n is the number of stages, the length of the sequence is one less than 2^n.) To wire up the circuit you could use a 74164 or 7474.

To deduce from the circuit diagram the sequence of numbers which are generated would be a substantial undertaking. We merely list the sequence here, using decimal number notation: 0, 32, 48, 56, 60, 62, 31, 47, 55, 59, 61,

[5]From *TTL Cookbook* by Don Lancaster, pages 278–79. Copyright © 1974 by Howard W. Sams & Co., Inc., Indianapolis, Indiana. Used with permission.

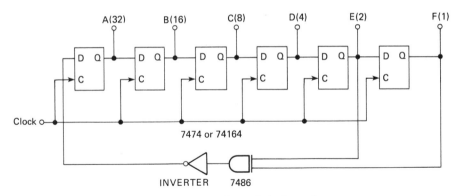

With inverter — State 111111 is disallowed; Series is complementary
Without inverter — State 000000 is disallowed; Series is normal.

Figure 13.8   A pseudorandom number generator.

30, 15, 39, 51, 57, 28, 46, 23, 43, 53, 26, 13, 6, 3, 33, 16, 40, 52, 58, 29, 14, 7, 35, 49, 24, 44, 54, 27, 45, 22, 11, 37, 18, 9, 4, 34, 17, 8, 36, 50, 25, 12, 38, 19, 41, 20, 42, 21, 10, 5, 2, 1. After the final 1, the sequence repeats, beginning with 0.

It appears to the eye that the numbers 0 through 62 occur well-mixed in the sequence. The extent to which the sequence satisfies rigorous mathematical requirements for true randomness is not considered here.

A peculiarity of the circuit shown is that the state 111111 (decimal 63) does not appear. If on power-up this should happen to turn up, the circuit will stay in that state. To prevent this eventuality, provision should be made to preset a 0 at some stage or to clear the whole register.

The pseudorandom generator can be used to generate random music. A circuit[6] for this purpose is given by Lancaster, while another use is in protection of computer data. The good computer data are mixed with the pseudorandom data to produce a jumbled string. To recover the good data, the pseudorandom elements in the string are removed, using the same sequence that was used in the scrambling process.

An Application in Computing

The role of shift registers in a computer can be nicely exhibited in the case of the multiplication of two numbers. Suppose that the product of the decimal numbers 36 and 21 is to be formed. Table 13.3(a) spells the process out. Lines

[6]Ibid., page 282.

A and B are the multiplicand and the multiplier. In line C 36 has been multiplied by the 1 in 21. In line D, 36 has been multiplied by the 2 in 21. In terms of hardware these operations would require some electronics to carry out each of the two multiplications, and to retain the results. In line E the partial product 36 is repeated, and in line F the partial product 72 has been shifted one place to the left. The need for a shift register appears here. Finally, the numbers in lines E and F are added column by column. The result is shown in line G.

Table 13.3 Multiplication of two numbers. (a) 36 times 21 in decimal notation. (b) 36 times 21 in binary notation.

A	36	a	100100
B	21	b	10101
C	36	c	100100
D	72	d	000000
		e	100100
		f	000000
E	36	g	100100
F	720		
		h	100100
G	756	i	000000
		j	100100
		k	000000
		l	100100
		m	1011110100

In Table 13.3(b) the corresponding steps are taken, but all in the binary number system. In line c an interesting fact emerges. A computer operating in binary can easily multiply any number by 1, for the product is simply the original number. In line d the result is also simple, for the product of any number and the number 0 is the number 0. In lines h through l shift-left operations have been performed. Finally in line m the sum of the numbers in lines h through l has been formed. In this sequence of operations the binary hardware needed includes left shift registers. The rest of the operations can be carried out by storage registers, full-adders, and subsystems which sense whether multiplication of a number by 1 or by 0 is required, and that in the first case merely reproduces the number n, and in the second case produce zero for the product. In programming a microcomputer or a larger computer at the symbolic language level considerable use is made of the ability to shift the contents of registers.

1. A shift register made from flip-flops

The essential workings of a shift register can be seen in stark relief by making one out of flip-flops. Figures 13.1 and 13.2 show how D flip-flops can be used to make an SISO and an SIPO shift register. The difference between the circuits is slight, so that both can be tried as easily as one.

2. TTL shift registers

You may wish to try wiring up a shift register circuit using any of the common TTL devices, such as the 74164, 74165, or 74195, and to get familiar with the operation of the circuit. Technical details concerning these are given in this chapter.

3. Serial-to-parallel conversion

The circuit shown in Figure 13.6 is a more sophisticated SIPO circuit, in that after the binary bits that were input *serially* are all stored, the entire set of bits can be transferred to another location in parallel fashion, on the arrival of a suitable command pulse. The way in which the circuit uses the clock pulses to control the storage and transfer of the data is especially interesting. This is done in a routine cyclic fashion, regulated pulse-by-pulse by the incoming clock signal.

4. A pseudorandom generator

The pseudorandom generator circuit shown in Figure 13.8 and discussed in the accompanying text is easily built and surprisingly interesting to study. The logic states of the six outputs can be monitored by LEDs. (Remember that limiting resistors may be needed to prevent burn-out of the LEDs. See the first section of Chapter 9.)

14

Some Other ICs

The 74181 Arithmetic Logic Unit

The 74181 Arithmetic Logic Unit (ALU) chip is phenomenally versatile in comparison with any we have met before. It is capable of carrying out 32 different arithmetic operations and 16 different logic operations. These include addition, subtraction, shifting, magnitude comparison, XOR, NAND, AND, OR, and NOR. Yet the device is only a 24-pin device. The pin diagram is shown in Figure 14.1.

The selection of the operation desired is achieved by using the inputs M, S0, S1, S2, and S3. Table 14.1[1] gives the possible operations as they correspond to the 32 different combinations of these five inputs.

In the notation used in the table, $\overline{A}$ is the complement of A; A + B means A OR B; A $\oplus$ means A XOR B; and AB means A AND B. The arithmetic sum of A and B is shown as A PLUS B. MINUS 1 following some symbol means that the two's complement of whatever is represented by that symbol is formed.

Each of the variables A, B, and F is a four-digit binary number. The digits of variables A are A3, A2, A1, A0, and similarly for B and F. The pins labeled C_n and C_{n+4} are the carry input and carry output pins, respectively. The A = B pin gives a comparator output. This pin will become high only when A is the same as B bit-by-bit. The pins labeled P and G are used when a full-carry look-ahead[2] system is wanted. They serve as outputs to be connected to the Look-Ahead Carry Generator 74182.

---

[1] From *TTL Data Book*, National Semiconductor Corporation, page 2–109. National Semiconductor Corporation. Copyright © 1981. Used with permission.

[2] "Look-ahead carry" is explained briefly in the third section of this chapter.

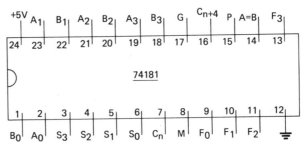

Figure 14.1 Pin diagram of the 74181 Arithmetic Logic Unit.

Table 14.1 Truth table for the 74181 Arithmetic Logic Unit

					ACTIVE LOW DATA		
SELECTION				$M = H$	$M = L$; ARITHMETIC OPERATIONS		
				LOGIC			
S3	S2	S1	S0	FUNCTIONS	$C_n = L$ (no carry)	$C_n = H$ (with carry)	
L	L	L	L	$F = \overline{A}$	$F = A$ MINUS 1	$F = A$	
L	L	L	H	$F = \overline{AB}$	$F = AB$ MINUS 1	$F = AB$	
L	L	H	L	$F = \overline{A} + B$	$F = A\overline{B}$ MINUS 1	$F = A\overline{B}$	
L	L	H	H	$F = 1$	$F = $ MINUS 1 (2's COMP)	$F = $ ZERO	
L	H	L	L	$F = \overline{A + B}$	$F = A$ PLUS $(A + \overline{B})$	$F = A$ PLUS $(A + \overline{B})$ PLUS 1	
L	H	L	H	$F = \overline{B}$	$F = AB$ PLUS $(A + \overline{B})$	$F = AB$ PLUS $(A + \overline{B})$ PLUS 1	
L	H	H	L	$F = \overline{A \oplus B}$	$F = A$ MINUS B MINUS 1	$F = A$ MINUS B	
L	H	H	H	$F = A + \overline{B}$	$F = A + \overline{B}$	$F = (A + \overline{B})$ PLUS 1	
H	L	L	L	$F = \overline{A}B$	$F = A$ PLUS $(A + B)$	$F = A$ PLUS $(A + B)$ PLUS 1	
H	L	L	H	$F = A \oplus B$	$F = A$ PLUS B	$F = A$ PLUS B PLUS 1	
H	L	H	L	$F = B$	$F = A\overline{B}$ PLUS $(A + B)$	$F = A\overline{B}$ PLUS $(A + B)$ PLUS 1	
H	L	H	H	$F = A + B$	$F = A + B$	$F = (A + B)$ PLUS 1	
H	H	L	L	$F = 0$	$F = A$ PLUS A*	$F = A$ PLUS A PLUS 1	
H	H	L	H	$F = A\overline{B}$	$F = AB$ PLUS A	$F = AB$ PLUS A PLUS 1	
H	H	H	L	$F = AB$	$F = A\overline{B}$ PLUS A	$F = A\overline{B}$ PLUS A PLUS 1	
H	H	H	H	$F = A$	$F = A$	$F = A$ PLUS 1	

*Each bit is shifted to the next more significant position.

The circuit shown in Figure 14.2 shows how two ALUs can be used to perform operations on *three* variables: A, B, and C. In the drawing the 74181s are wired up to perform the operation $(A \oplus B) \oplus C$. This function could be quite easily implemented with a few NOR gates, and it might be argued that the circuit shown is not an improvement over the NOR gate circuit. You can show that the complexity of the wiring interconnections is about the same in either case. However, the versatility of the ALU circuit offers an enormous

150

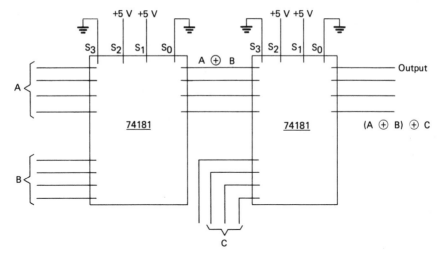

Figure 14.2   Use of two arithmetic logic units to perform a special logic function.

advantage over the fixed-purpose NOR gate circuit. By simply changing the selection inputs (S) the ALU circuit can be made to perform many different operations. A few examples are (A + B)C, ABC, and (AB)C. A different system using basic logic gates would have to be built in each of those cases and switching from one to another would be a substantial problem.

A TTL Adder

In Chapter 3 where we discussed adders, we saw how they can be constructed with logic gates. As you would expect, full-adders are available in TTL packages. One is the 7483, the pin diagram for which is shown in Figure 14.3.

This device produces the sum of two binary numbers, each made up of four binary digits. The digits of one of the addends (A) is input at pins 1, 3, 8, and 10, with corresponding weights 8, 4, 2, and 1. The digits of the other addend (B) are input at pins 16, 4, 7, and 11 with corresponding weights 8, 4, 2, and 1. The sum (S) appears at pins 15, 2, 6, and 9 with weighting 8, 4, 2, and 1. If the sum is not greater than 1111 (decimal 15) a 0 appears at the carry-out pin 14 (marked C4), otherwise a 1 appears there. Pin 13 (marked C0) should be grounded.

Two 7483s can be used to make possible the addition of two 8-digit numbers. The carry output from pin 14 of the lower-order adder is connected to pin 13 (C0) of the higher-order adder in order to supply the latter with the carry information needed to complete the addition.

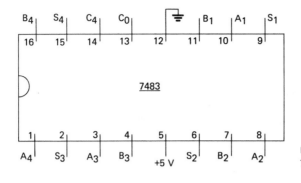

Figure 14.3 Pin diagram of the 7483 adder.

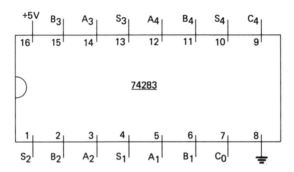

Figure 14.4 Pin diagram of the 74283 adder.

The 74283 is a similar full-adder. The pin diagram is shown in Figure 14.4. It differs from the 7483 in this respect: While the carry from one 7483 to another (as in the addition of two 8-digit numbers) is made in ripple fashion, the coupling of two 74283s provides a faster way to deal with the carry bit. This method is known as *look-ahead carry*. If the half-adders of Figure 3.1 are used to make a four-bit adder, there is a relatively long time involved in propagating signals through the circuit. It has been found possible to change the arrangement of gates in the four-bit adder so that the propagation time is reduced by a short cut, which is the look-ahead feature.

Other Arithmetic Operations

It is interesting to observe how the useful functions of subtraction and multiplication[3] can be carried out by using only an adder chip and a few gates. Figure 14.5 is a subtractor circuit[4] which can form the difference $(A - B)$ where each of A and B is a 4-digit number.

---

[3]Multiplier chips are available, but are relatively expensive. Division (as in microcomputers) is usually carried out by software (special programs).

[4]Based on material from *Introduction to Digital Techniques* by D.I. Porat and A. Barna, page 340. John Wiley and Sons, Inc. Copyright © 1979. Used with permission.

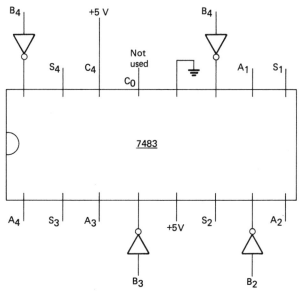

Figure 14.5 Conversion of a 7483 adder into a subtractor.

As is explained in Appendix A, the difference (A − B) can be formed by adding to A the two's complement of B. The two's complement of B is formed by taking the one's complement of B and adding 1 to it. In the circuit shown the inverters form the one's complement of B, and the 1 at the carry-in pin completes the production of the two's complement of B.

The circuit[5] of Figure 14.6 produces the 6-digit product P of a 4-digit number X and a two-digit multiplier Y. The circuit uses a 7483 and eight AND gates. Although the circuit is shown in unusual piecemeal fashion in parts (a), (b), and (c) of Figure 14.6, this will make the following discussion easier to follow. First notice that if X is multiplied by the number 1, the product is identical with X, and that if X is multiplied by the number 0, the product is 0. Thus the multiplication of X by either a 1 or a 0 is simple. To multiply our X by the 2-digit Y, the two digits of Y are used to form the *partial products,* each of which is X or 0; the second of the partial products is shifted one place to the left; and the sum of the first partial product and the shifted second partial product is formed. For example, 1101 times 10 is the sum of the partial product 0000 and the shifted partial product 11010, or 11010.

In Figure 14.6(b) of the circuit the number X is multiplied by the digit Y1 by an independent gate. This forms the P1 digit of the product off the chip, and is a trick by which the relative shift of the two partial products with respect to each other is accomplished. The remaining digits P2, P3, P4, P5, and P6 of the product appear at pins 9, 6, 2, 15, and 14 of the 7483.

[5]Ibid., page 349.

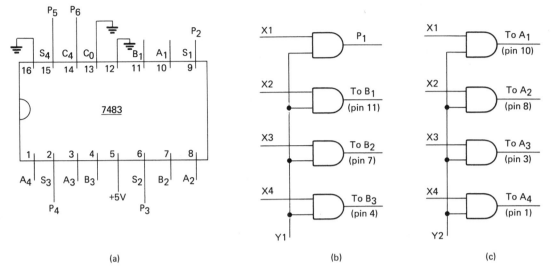

(a) (b) (c)

Figure 14.6 Conversion of a 7483 adder into a multiplier.

Digital Comparator

A *digital comparator* is a device which compares two binary numbers and at an output reports whether the numbers are equal, or which is the larger if they are unequal. The 7485 is a frequently used digital comparator. The pin diagram is shown in Figure 14.7.

Figure 14.7 (a) Pin diagram of the 7485 digital comparator; (b) truth table for the 7485.

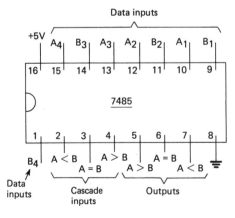

	Pin 5	Pin 6	Pin 7
$A > B$	1	0	0
$A = B$	0	1	0
$A < B$	0	0	1

(a) (b)

If one 7485 is being used to compare two 4-digit numbers A and B, the *cascade input* pins must be connected with A > B and A < B to ground, and with A = B to +5 V. Then data output pins, 5, 6, and 7 respond according to Figure 14.7(b). There is in any case one and only one output pin high.

Two 7485s can be cascaded to make possible the comparison of two 8-digit binary numbers. To do this the output pins of the comparator which handles the low-order digits are connected to the cascade inputs of the comparator which handles the high-order digits. The output then appears at the output pins of the high-order comparator.

A digital comparator could be used to count the number of times some particular number X occurs in a long string of numbers. The reference number X would be supplied to the comparator, and the numbers in the string would be applied one at a time to the other inputs of the comparator. A 1 would appear at the A = B output pin whenever an X appears in the string.

A comparator could be the basis of a system which would pick out the largest number from a set of numbers. Let the numbers be called A, B, C, D, and so on. Let A and B be input to the comparator, and let a B > A output signal be used to cause some associated circuitry to hold B as input to the comparator and to cause C to become the other input in place of A. In the next step, the larger of B and C would be held, and so on. There would have to be provision (not discussed here) for avoiding ambiguity in any case in which the inputs to the comparator are equal. By an extension of this basic idea and with some more circuitry, a string of numbers could be sorted into numerical order.

We offer a final example. Suppose a digital system contains a subtractor which is such that it can produce the difference $(A - B)$ if B is equal to A or less than A. The capability of the system can be extended so that it can form the difference $(A - B)$ in the remaining case in which B is greater than A. In order to do this, A and B would first be compared by a comparator, and if the result is A = B or A > B the subtractor would be authorized to form $(A - B)$, but if the result is A < B the roles of A and B as inputs to the subtractor would be reversed, the difference would be formed, and finally the sign of the result would be changed.

Priority Encoder

In this section we discuss the 74148 Eight-Bit Priority Encoder. The nature of other priority encoders is similar. The 74148 outputs the location of the highest-ranking of eight input pins which is at logic 0. For example, if the input digits are 11001011 then the output will be 010 (reading from right to left, as is usual). Again, for input 11101111 the output would be 011. The pin diagram for the 74148 is shown in Figure 14.8.

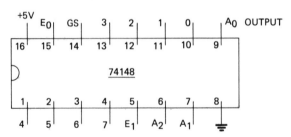

Figure 14.8 (a) Pin diagram of the 74148 priority encoder.

In addition to the expected eight input pins, three output pins, and two power supply pins, there are two additional output pins and one additional input pin. Output E0 and input E1 are used as follows: Input E1 must be low to have outputs while the E0 output will go high if the outputs are activated. These two pins are used for cascading. Output GS goes low if any input is selected and the enable E1 is low.

A prime use for priority encoders is in microcomputer systems. Such systems usually have a number of *input ports,* which communicate with various input devices, such as teletypes and instrument sensors. By a simple sequencing operation, the microcomputer looks at its input ports over and over while it is carrying out its normal functioning, such as computing according to some main program. If in the sequencing the microcomputer perceives that one of the external devices signals at an input port that it wants attention, the microcomputer goes into an *interrupt service routine,* in which it breaks off its normal operation and does whatever it is that the external device requires, and after that returns to its normal computing and scanning of the input ports. The problem which arises if two or more of the external devices demand attention simultaneously can be handled by assigning priorities to the devices and by using a priority encoder to rank the input signals accordingly.

The priority encoder also finds uses in code converters and generators. The 74148 can be regarded as an eight-line to three-line encoder. A cousin of the 74148 is the 74147, which is a priority encoder which can be called a nine-line to four-line BCD encoder. The circuit of Figure 14.9 shows schematically how the 74147 can be used with a nine-key keyboard to allow an operator of the keyboard to show on a seven-segment display which key he or she has depressed and simultaneously to transmit that information to an electronic system for further processing, such as some calculating and some printing. Such a system might be used by a teller at a parimutuel window. The encoder translates the information which is input when a key is pressed into BCD format, and after the BCD information is input to "memory" it is available for any processing which is required, while the BCD-to-seven-segment decoder

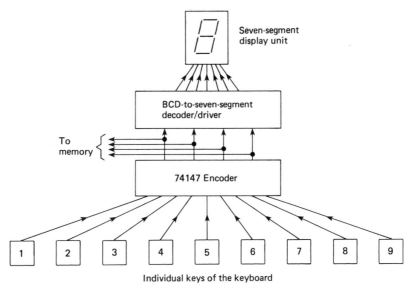

Figure 14.9 The use of a priority encoder.

shows on the seven-segment display unit which key was depressed. If no key is depressed, the BCD output is automatically 000, and the display unit would exhibit 0.

A Sampling of Still Other IC Devices

The richness of varieties of IC devices which are available makes anything approaching completeness in a short listing impossible. However, it is worthwhile for a student of ICs to get some idea of what can be had today. For that reason we offer in this section an unashamedly arbitrary sampling. This will be only a short ramble through a few parts of the world of ICs where there are enormously important devices (CPUs) and numerous more obscure but important—when an application needs them!—devices.

Because many digital systems today have keyboards by which the user inputs data, *keyboard encoders* are common. When a key is depressed, the output pins of the device show a four-bit binary equivalent which will be recognized as meaningful by the device with which the keyboard communicates.

The prevalence of cathode-ray tube monitors in association with computers has led to the mass-production of *character generators*. In one case, the character generator accepts ASCII-coded information (see Appendix B) and produces alphanumeric information (letters, decimal digits, and some other symbols) in form compatible with the 5-by-7 matrix display on a line-by-line scan of a display screen.

Calculators and other applications require a bank of seven-segment display units, decoders, latches, and multiplexers. You can now get off-the-dealer's-shelf items which contain some or even all of those things, all compactly built into small packages. Counters with built-in latches are simple examples.

The popularity of digital clocks with or without alarms, various alarm options such as snooze and radio turn-on features, has led to the production of remarkably low-cost packages. Such a package may be smaller than 2 inches by 3 or 4 inches, including the display. When such a clock device is to be operated from the 110-volt AC lines, some more hardware is needed to convert the 60-hertz 110-volt line signal into a form usable as the seconds signal by the clock module. This requires rectifier diodes, capacitors, and other elements external to the chip itself. You can get complete *clock modules* at low cost which have on a small board everything needed to go into operation except a special transformer and the control switches. Such a combination of a digital device with such components as capacitors in a single unit makes up a *hybrid* device.

An entire (and large) book would be required to discuss the *central processing units* (CPUs), which are the cores of computer systems. It is a tribute to human ingenuity and the capabilities of modern technology that extremely capable CPUs can be bought for as little as twenty or so dollars today. What a small 48-pin CPU is able to do in association with a handful of other chips is truly remarkable. The spread of microcomputers used both as computers and as dedicated control devices throughout the world constitutes an electronic revolution in itself.

Even in the restricted domain of relatively simple combinatorial devices the variety of available ICs is large and growing. For example, there are the Dual Two-Wide AND-OR-INVERT Gate (7450) and related ICs. There are also new and interesting *majority* gates, which have multiple inputs and an output that indicates which inputs were chosen in the majority of cases.

EXPERIMENTS

1. Familiarization with the arithmetic logic unit

A very interesting—and nontrivial—undertaking would be to verify by experiment with the chip itself that you can make an ALU live up to the potentialities promised. In this chapter we discuss the 74181 as an example of an ALU. The truth table (Table 14.1) is rich with possibilities.

2. A subtractor circuit based on a TTL adder chip

In the third section of this chapter we discuss the conversion of the 7483 adder into a subtractor with the use of some external circuitry. This is a worthwhile circuit to consider conceptually and to build with actual hardware. The essentials of the circuit are shown in Figure 14.5.

3. A multiplier circuit based on a TTL adder chip

In present-day microcomputer technology, microprocessor chips are capable of addition and subtraction as well as of many other operations directly, by virtue of their inner hard-wired circuitry, but multiplication and division must be carried out by programming *software* by which the chips multiply and divide by addition, subtraction, and the other operations the chips are capable of directly, such as shift operations and move-from-here-to-there operations. The circuit shown in Figure 14.6 is interesting in this context. It is a circuit to convert an adder chip into a multiplier with some external circuitry. (We must add that at the time of this writing, microprocessor chips with the ability to carry out all of the operations, $+$, $-$, $\times$, and $\div$ are coming onto the market.)

4. Other TTL devices

In the last three sections of this chapter we call attention in very cursory fashion to some of an increasing number of special TTL devices. According to your particular interests, you may want to learn more about some of these by wiring them up and getting familiar with them.

5. Advanced design techniques

In connection with Experiment 2 in Chapter 12 we made some comments about using an IC device to accomplish some end which the name of the device itself does not suggest. For example, a multiplexer can be used to generate a truth table of a nonstandard kind. A RAM can be used to generate quite elaborate truth tables. In both of those cases, the circuit that results is simple in concept and in practice.

In the present chapter some other ICs are mentioned. Each of these has a name that indicates what the manufacturer considered to be its primary use, such as *keyboard encoder* or *digital comparator*. However, it is quite possible that some of these miscellaneous devices can be adapted to do something unusual, or to do something easily that would require complicated circuitry otherwise. We suggest that you bear the possibility in mind. Something interesting might turn up. *Popular Electronics, Radio-Electronics,* and other magazines welcome novel circuit ideas from their readers.

CMOS and
Other Logic Families

Introduction to Logic Families

The TTL (transistor-transistor logic) integrated circuits we have discussed so far in this book consist of NPN transistors (which we will describe in detail in the third section of this chapter) interconnected in certain ways. In DTL, or diode-transistor logic, the basic elements are both diodes and transistors. Such distinctions specify the different *logic families*.

The most widely used logic families today are TTL and CMOS. Some of the other families were developed relatively long ago when the digital integrated circuit era was dawning and still others are of quite recent origin. While logic families other than TTL and CMOS are used comparatively rarely, you should be aware in at least a general way of their names and the ways in which they differ with respect to their principal properties. For this reason, in the next section of this chapter we will survey briefly the most significant logic families.

In later sections we will discuss the NPN transistors which TTL and some other logic families use and the MOS elements which are used in MOS logic. This is done to help you understand the remaining sections of the chapter where CMOS is discussed in some detail.

A Survey of the Major Logic Families

This section will consist of a list of the main logic families in use, with no attempt at more than a cursory indication of the ways in which they differ. We

160

must, of necessity, refer the reader who is interested in finer details to the more specialized literature.

The names of the first four families listed are descriptive of the internal natures of the devices. For example, in RTL (resistor-transistor logic) the gates and other functional blocks are built of resistors and transistors. In ECL (the fifth family), the internal transistors are wired together pairwise with their emitters coupled together, justifying the name "emitter-coupled logic." In IIL or I^2L (the sixth family), pairs of transistors are coupled in still another way to make gates and other functional blocks. The remaining two families (MOS and CMOS) will be handled in later sections of this chapter.

TTL or transistor-transistor logic

Comments: Cheap; readily available with many functions; easy to use; popular. Widely known and used.

Power consumption: High

Speed: Fast

Noise immunity: Good

DTL or diode-transistor logic

Comments: Cheap; easy to use. Not commonly used.

Power consumption: High

Speed: Fast

Noise immunity: Good

RTL or resistor-transistor logic

Comments: Cheap. Not common.

Power consumption: Low

Speed: Slow

Noise immunity: Very good

RCTL or resistor-capacitor-transistor logic

Comments: It is difficult in manufacture to make compact capacitors[1] in ICs. Therefore, RCTL cannot have a high density of components. Little used.

Power consumption: High

Speed: Fast

Noise immunity: Poor

ECL or emitter-coupled logic

Comments: ECL provides the highest speeds at present; requires more than one power supply; not easily interfaced with other families. Not common.

---

[1]Aside from such active circuit elements as transistors, electrical circuits are made of resistors, capacitors, and inductors. It is interesting to observe that inductors are not built into ICs at all.

Power consumption: High
Speed: Very fast
Noise immunity: Poor

IIL (I²L) or integrated injection logic

Comments: High internal density, conducive to sophisticated devices; easy
to interface. Not very familiar to most users of ICs.
Power consumption: Low
Speed: Slow
Noise immunity: Good

MOS or metal-oxide-semiconductor logic

Comments: Capable of very high internal density; much used in specialized devices such as digital watch and clock chips; cheap,
even when sophisticated.
Power consumption: Extremely low
Speed: Slower than TTL
Noise immunity: Very good

CMOS or complementary metal-oxide-semiconductor logic

Comments: High internal density; tolerant of a range of power supply
voltages; easy to interface with other devices; available in a
wide range of functions. Very popular.
Power consumption: Fantastically low
Speed: Slower than TTL, generally speaking
Noise immunity: Very good

The Nature of the NPN Transistor

In this section and the next, the basic building blocks of TTL and MOS are
described. The goal sought is a general understanding of the properties which
make the two contrast so highly. A spinoff will be a better understanding of
CMOS, discussed in following sections.

Figure 15.1(a) shows in schematic form an ordinary NPN *bipolar* transistor. It consists of a layer of P-type semiconductor material sandwiched between layers of N-type semiconductor material. The interface between the
emitter and base layers constitutes a diode, as represented in Figure 15.1(b).[2]
The diode conducts only when the base is made positive with respect to the
emitter. In part (c) the external circuit from the emitter to the collector
includes a voltage source and a current-limiting load resistor R_L.

[2]See the Discrete LED section of Chapter 9 for more about diodes.

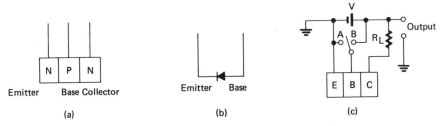

Figure 15.1   Some basic NPN transistor concepts.

Unless the emitter-base diode is turned on by a suitable potential difference between the emitter and base sections, conduction through the
transistor cannot occur. In part (c) there is a switch which is installed in such a
way that if it is thrown to position A, the emitter-base diode is OFF and current
does not flow through the external circuit, but if the switch is thrown to
position B, the diode goes ON and current flows in the external circuit.

By throwing the switch alternately from A to B to A and so on, the voltage
between the output terminals can be made to alternate between high and low
levels. Hence a transistor used in this way is a device with two output states
which can be identified with the logic states 1 and 0. It is this property which
makes this kind of transistor of interest in digital ICs.

The manufacture of ICs based on NPN transistors is both technically and
economically feasable. The widespread use of TTL digital ICs testifies to this.

The Nature of MOS Devices

Figure 15.2(a) shows the structure of a related device which is, nevertheless,
quite different. This is known as an *enhancement mode metal-oxide-semiconductor*

Figure 15.2   Some basic MOS concepts.

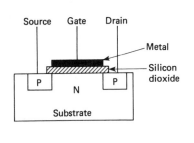

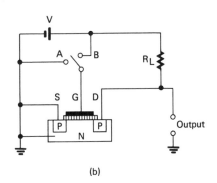

(a)

(b)

transistor, or more briefly *metal-oxide-semiconductor* transistor, MOS transistor, or simply MOS.

The MOS shown has two P-type segments inserted into an N-type underlying slab which is called the *substrate*. The wire lead to the leftmost P-type segment is called the *source* lead and the wire lead to the rightmost P-type segment is called the *drain* lead. Notice that where the transistor in Figure 15.1 has a wire lead making direct conducting or *ohmic* contact with the N-type substrate, the gate lead of the MOS goes to a metal film which is separated from the main body of the device by an insulating layer. The insulating layer is usually made of silicon dioxide (glass) although other substances have been used.

In order to compare the behavior of this MOS *monopolar* transistor with that of an NPN bipolar transistor in a comparable circuit, we show Figure 15.2(b). If the manual switch is thrown to position A there is no conduction through the device from source to drain. This is because the diode junctions at the interfaces of the source and drain segments with the substrate are not both biased on. If the manual switch is thrown to position B, a conducting channel is formed between the source and drain segments through the substrate in the region just under the glass insulating layer. In short, with the switch at A current does not flow through the device and the external circuit and with the switch at B, the current does flow.

Again we can associate the output voltage levels with the logic states 1 and 0. When the switch is at A, the output is a positive voltage and when the switch is at B, the output is at zero volts, both with respect to the circuit ground. (Due to a small leakage current through the MOS device, the voltage levels at the output are actually other than +V and 0 by small amounts, a few millivolts at most.)

Both the logical behavior of the circuit and its most important electrical features can be represented in a very simple way. This is shown in Figure 15.3. In part (a) the signal at the gate is at logic 0 and the MOS device itself is

Figure 15.3   The single MOS, as a switching device.

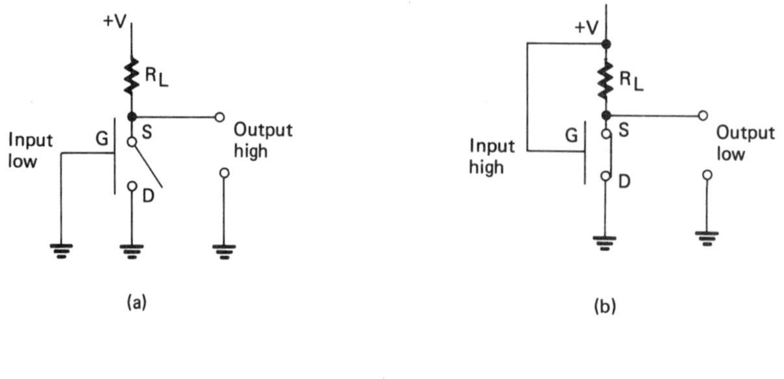

(a)

(b)

nonconducting between source and drain. This is indicated by showing an open switch between source and drain. (This switch is not to be confused with the manual switch in Figure 15.2). In part (a) the output terminal runs directly to the +V terminal of the voltage source with no voltage drop across resistor R_L, since no current flows through R_L. Therefore the voltage level at the output terminal is +V (within a few millivolts) and we can take the level there to be 1. In part (b) the gate lead is connected to the +V terminal and so is at logic level 1 and the output terminal runs directly (or nearly so) to ground and can be said to be at logic 0. Hence this MOS device is a simple inverter.

The MOS shown in Figure 15.2 is an *N-channel MOS* or *N-MOS*. If you reverse the labelings of the N and P segments of the sketch, you will arrive at a *P-channel MOS* or *P-MOS*. Where in Figure 15.3(a) the logic low input at the gate makes the source-to-drain switch open for N-MOS, a logic low input at the gate makes the switch close in P-MOS. In Figure 15.3(b) the logic high gate makes the N-MOS switch close but would make a P-MOS switch open. For clarity, these features of the two kinds of MOS circuits are collected in Figures 15.4 and 15.5. In Figure 15.4 we have an N-MOS and in Figure 15.5 a P-MOS.

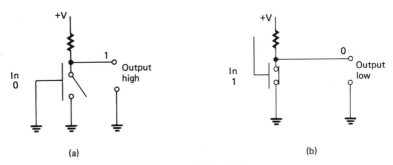

(a)   (b)

Figure 15.4   The single N-MOS, as a switching device.

Figure 15.5   The single P-MOS, as a switching device.

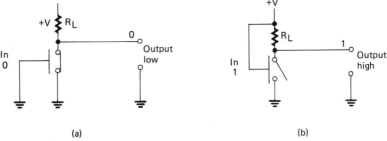

(a)   (b)

In Figure 15.4 (a) and (b) you can see that the rule "in low, out high; in high, out low" applies, hence the circuit is that of an inverter. In Figure 15.5 (a) and (b), you can see that the rule "in low, out low; in high, out high" applies. This circuit is a noninverting buffer.

There is a feature of MOS devices which is of the utmost importance. In order to set the state of the gate of an MOS device, some signal source must bring the gate high or low. However, since the gate lead is insulated from the channel of the MOS device, there will be no continuous current flow between the signal source and the MOS device proper. The signal source must handle some *transient* current to charge or discharge the capacitor which consists of the gate metal plate, the insulating glass layer, and the channel itself. However, this state-switching current is very minute. Hence we can say that driving an MOS device demands virtually zero power from the signal source in standby conditions and extremely little power under state-switching conditions. But notice that in Figures 15.4 (a) and 15.5 (a) there *are* conducting paths from the power supply to the circuit ground.

The Nature of CMOS

The basic CMOS unit consists of a pair of MOS transistors. One is N-MOS and the other P-MOS. The pair is said to be complementary and is known as a *CMOS* unit. The name CMOS is derived from *complementary metal-oxide-semiconductor*.

The manner in which the MOS devices are arranged in a CMOS unit is shown in Figure 15.6. The input lead to the device is at logic 1 in part (a). The output lead is disconnected from +V and connected to the circuit ground. We have here "input high, output low." In part (b) the input lead to the CMOS

Figure 15.6   How the CMOS (complementary MOS) pair works.

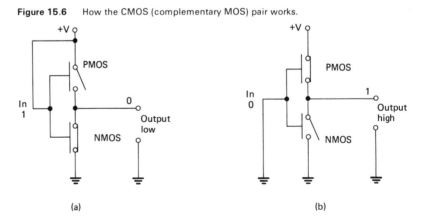

(a)   (b)

device is at logic 0. The output lead is connected to +V and disconnected from the circuit ground. We now have "input low, output high." The CMOS pair is thus an inverter.

General Discussion of CMOS

The CMOS unit has the important property that it demands no current[3] from the signal source when the gate level is constant and very little current from the signal source when the gate level is being switched. But also notice that whether the output state is 0 or 1, there is no current path from the +V terminal of the power supply through the MOS channels to the circuit ground. Thus the CMOS unit expends virtually no power in its standby states and very little power when in operation. This is a very attractive feature of CMOS. For example, in a battery-driven CMOS device, the battery can last almost as long in use as it would if it were merely sitting on a shelf. In contrast, electronic devices which are based on TTL ICs make much more substantial power demands on the supplies which drive them.

How this works out numerically is striking. Since the signal source is required to drive very little current the input impedance of CMOS is very high. In fact, values in excess of 10^{12} ohms are usual. For an assumed input signal of 10 volts, the Ohm's law relation $I = V/R$ gives us a corresponding value of the *signal* current only 10^{-11} amperes. The corresponding signal source power is $P = IV = 10^{-10}$ watts, or 1/10 nanowatt. For a CMOS chip the *overall* power consumption (standby included) is of the order of 10 nanowatts. For a voltage of 10 volts, the relation $I = P/V$ leads to an overall average current of only 10^{-9} amperes.

We would expect the fanout to be very large for CMOS in view of the characteristics mentioned. A numerical value of 50 for the fanout is often cited in the literature. Fanout clearly becomes a problem only in very large and complex systems.

When any IC (whether CMOS or not) is a source which drives some load, the output impedance of the IC can be important. Almost always you want the output impedance of a source to be low. Again CMOS offers a distinct advantage in that the output impedance is indeed low. A typical value is 500 ohms or so whether the logic state of the output pin is 0 or 1.

CMOS can be powered by voltages with much relaxed specifications in comparison with TTL. The voltage can range from about 3 to about 18 volts.[4] Notice that this range includes 5 volts. This means that a single voltage

---

[3] In speaking of "no current" a slight approximation is made because of the existence of minute leakage currents.

[4] MOS and CMOS devices sometimes use unusual voltages, for example 1.5 volts for chips used in digital watches.

source could be used to power a system in which there are both TTL and CMOS chips. (However, interfacing problems may have to be solved in such a system. This is discussed briefly further on.) Bear in mind that both the maximum operating speed and the power expended depend on the voltage used. The maximum speed is quite sensitive to the voltage and the manufacturer's data should be consulted if this is likely to be of concern. Generally speaking, 10 to 12 volts is best. Yet another advantage is that the power supply need not be tightly regulated. You would expect this, given the tolerance of CMOS for a wide range of voltages.

CMOS devices are either *buffered* or *unbuffered*. The unbuffered devices use the basic unit shown in Figure 15.5. Buffered CMOS uses a basic unit followed by additional basic units, each of which is driven by the preceding unit. A chief difference between the behaviors of the two kinds is shown in Figure 15.7. The noise immunity of the buffered variety is substantially better than for the unbuffered variety. In the latter case, noise on the line can more easily bring the device into the transition region between definite logic 1 and definite logic 0.

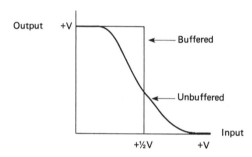

Figure 15.7   The input-output characteristics of buffered and unbuffered CMOS.

Also, as the sketch indicates, CMOS devices swing all the way from zero volts to +V volts when they switch from output low to output high.[5] TTL devices can be considerably above zero volts when at logic low and considerably below +5 volts when at logic high.

It is often said that a disadvantage of CMOS is that it has low-speed operation. It is true that CMOS at its fastest does not rival TTL at *its* fastest, but CMOS can be used at frequencies up to several megahertz and propagation times are often from about 50 to nanoseconds. Thus CMOS is fast enough for many applications even if it is not extremely fast.

The maximum and minimum operating temperatures for regular com-

---

[5]This is true to within a few millivolts at most. The contrast with TTL justifies the small exageration made in the statement.

mercial CMOS chips are 85°C (185°F) and −40°C (−40°F). Improvement can be achieved by using military specification grade CMOS chips for which the corresponding temperatures are 125°C (257°F) and −55°C (−67°F).

Some Further Remarks About CMOS

A significant shortcoming of CMOS is that it can easily be destroyed by static electricity. When you are installing CMOS in a circuit or even when you are just picking a CMOS package up, you may deliver a disastrous voltage between two pins unless precautions are taken. When you walk on a floor, brush your hand across a workbench or touch your hair, you can take on a potential of thousands of volts, especially under conditions of low humidity. If your fingers touch the pins of a CMOS device when you are in such an electrical state, the result can be rupture of the silicon dioxide layer somewhere in the device.

Most CMOS chips are now protected by built-in diodes or other circuitry which are meant to bypass dangerous current around the MOS units themselves and to the substrate. However, this does not provide complete protection.

Basically what you must do to avoid damaging CMOS devices is to make it impossible for a potential difference to exist between pins when handling them. A stored CMOS is safe if it is kept with its pins stuck into a piece of *conductive* pastic. Neither a CMOS package nor a circuit which contains CMOS ICs should be put into any other kind of plastic container. When you must handle CMOS, you should take steps to avoid touching it with a part of your body or with a tool which may be at a high potential. Wearing a conductive wrist strap connected through a high resistance to ground or similarly grounding the tool is recommended. This is a rather elaborate precaution if you are only gong to handle one CMOS long enough to insert it into a circuit. In such a case, you can touch a water pipe or other really good ground connection immediately before handling the device. Some other rules are not to handle ICs by their pins, not to put a chip down once you have picked it up and not to hand a chip to someone else who is not protected as you are.

We turn now to another simple but essential point about using CMOS devices. With TTL an input pin that is left unconnected will act as if it were at logic high. This rule does not apply to CMOS. With CMOS every input pin must be connected with to +V, to the circuit ground, or to the appropriate output pin of some chip.

Finally we offer some remarks about mixing CMOS and TTL devices, while not attempting to deal with the interfacing problems which can arise in anything more than a brief suggestive fashion. TTL requires +5 volts from its power supply, to within fairly tight limits, but CMOS will accept a rather wide range of power supply voltages. If a circuit contains both CMO and TTL and is

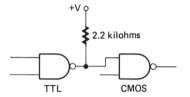

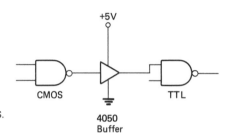

Figure 15.8   Driving CMOS with TTL.

Figure 15.9   Driving TTL with CMOS.

run from a single +5-volt supply, the interfacing problem is simplified. More generally, problems arise. Figure 15.8 shows a TTL gate driving a CMOS gate to give you an idea of a technique which may be used. The 2.2 kilohm resistor is known as a *pull-up* resistor. Its function is to bring the logic high state of the TTL gate more nearly up to the voltage of the power supply than would otherwise be the case.

On the other hand, driving TTL with CMOS is somewhat more complicated. A kind of CMOS gate called *series A* can drive any single TTL gate input, but a *series B* CMOS[6] gate output can drive only a low power TTL gate (LS or L series). To improve the fanout you need a suitable driver.

If the constraints are also that the CMOS chips must be operated with some voltage other than +5 volts, the problem becomes in part one of matching voltage levels. For example, in Figure 15.9 the CMOS gate is assumed to have an output level higher than 5 volts and the 4050 buffer is connected to the power supply which powers the TTL gate. The special 4050 device effects the conversion needed.

Concluding Remarks About MOS, CMOS, and TTL

The logic of TTL gates, counters, flip-flops, and many other devices carries over directly to the realm of CMOS devices. A practitioner who understands

170

[6]The B in series B suggests the word "buffered," but not all series B devices are buffered.

one family understands much about the other family. Mostly, you need to be aware of the differences. Some of the differences between the families have been discussed in the earlier parts of this chapter. In this section of the chapter, some further comments about differences (and similarities) are made.

It is possible in the manufacture of integrated circuit packages to pack many more MOS units into a given chip area than is true for TTL. Hence the more elaborate and sophisticated devices tend to be MOS. As many as 100,000 transistors on a chip are possible now, using MOS technology. Large memories and microprocessors are examples. Remarkably, even such powerful devices are quite low in cost.

Many CMOS devices are available which do not exist in TTL form but, on the other hand, many devices are available in both CMOS and TTL forms. In some cases the pin diagrams for a device are the same for the CMOS and TTL versions. In any such case, the devices are pin-for-pin interchangeable in a circuit provided power supply and interfacing problems do not intervene. In other cases, the pin diagrams may differ. For example, the TTL 7400 Quad Two-Input NAND and the CMOS 4011 Quad Two-Input NAND Gate are compared in Figure 15.10.

The numbering system is different from MOS and CMOS than for TTL. Unfortunately the matter is in a confused state at present. While there are fairly simple CMOS series such as the 4000 SII series, 4500 MSI series, and 5000 LSI series, there are many other numbering schemes which do not conform to these patterns. There is a slight trend toward adapting the TTL 7400 numbering system to CMOS so that (for example) a 74C00 is a CMOS version of the TTL 7400. However, you must consult the manufacturers' data books and other technical literature for help in associating chip numbers with their functions and characteristics.

Figure 15.10   Comparison of the pin diagrams of the TTL 7400 and the CMOS 4016 quad NAND gate ICs.

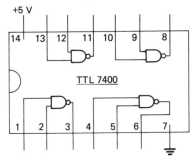
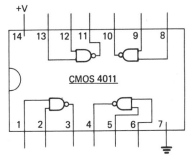

The NPN transistors in TTL integrated circuit chips and the complementary MOS pairs in CMOS chips are both switching devices. Each has a low state corresponding to binary 0 and a high state corresponding to binary 1. It follows that the *logic* of a circuit does not depend on whether TTL or CMOS is used. You should be able to build any of the circuits described in this book with CMOS, provided the CMOS equivalent of the TTL device exists. The same +5-volt power supply and all the other hardware will serve in either case. Often CMOS and TTL devices are pin-for-pin equivalent.

You will learn that CMOS and TTL are much alike, and that you do not necessarily ruin a CMOS chip merely by picking it up, for example. Mark Twain said that a man who tries to carry a cat home by the tail learns more than a man who does not try it.

It is with respect to their technical properties that TTL and CMOS differ, rather than with respect to their logical properties. The more technically inclined reader might choose to study the behavior of CMOS when various power supply voltages are used, the power consumption of CMOS, or interfacing with TTL, to mention a few of many possibilities. The computer-oriented experimenter might well want to study some large CMOS memory device. It is a rich field. Whichever way you, our reader, chooses to go, we wish you enjoyment!

Appendix A:
The Binary Number System

Each of the input and output pins of a digital integrated circuit device can be at one or the other of two voltage levels.[1] It is also the case that in the binary number system there are precisely two digits or *bits*: 0 and 1. Thus we can associate each of the pin states of an IC with one of the bits 0 or 1. The decimal number system uses 10 digits (0 through 9), and an electronic device would have to operate with 10 voltage levels for a similar association between voltage levels and digits to be made. Electronic circuitry based on the decimal number system would present formidable technical problems. In the real world today devices which switch between only two levels are used. These include digital ICs, magnetic tapes and discs, punched computer cards, and other hardware. Therefore, the binary number system is of basic importance in digital electronics. Sheer familiarity from childhood with the decimal number system gives you the feeling that it is simple. However, the binary number system is even simpler in important ways, as you will see.

How to Interpret a Given Binary Number

Consider first how you interpret a number which is given in decimal form. Take as an example the number 573. It is understood that the digit 5 means 5 hundreds, or 5×10^2; that the digit 7 means 7 tens, or 7×10^1; and that the digit 3 means 3 ones, or 3×10^0. Finally it is understood that the symbol 573 means that the number represented is the sum $(5 \times 10^2 + 7 \times 10^1 + 3 \times 10^0)$.

[1]Three-state logic (Chapter 4) is an exception.

Now consider how the binary number 1101 which consists entirely of the bits 0 and 1 is interpreted. The leftmost bit 1 is taken to mean 1×2^3, the following 1 to mean 1×2^2, the zero to mean 0×2^1 and the final 1 to mean 1×2^0. Finally, the number represented by 1101 is the sum of these quantities. Thus the number represented is statable in decimal notation as being $(8 + 4 + 0 + 1)$, which is equal to 13. Hence the binary number 1101 is a representation of the decimal number 13.

The difference between the binary and decimal schemes is that in the former the *weights* attached to the bits are powers of 2, while in the latter the weights attached to the digits are powers of 10.

In the longer chain of binary bits 1010110, the weights (reading now from right to left) are 1, 2, 4, 8, 16, 32, and 64. The decimal equivalent of the binary number given is 86.

It is very often necessary to be able to refer to the bits in a binary number in a general fashion. This is usually done by calling them A, B, C, D, E, and so on, reading from right to left. Thus EDCBA refers to a 5-bit binary number without specifying numerically what any one of the bits is.

In working with digital circuits you usually have to interpret rather short "words" which, in fact, turn out most often to be four-bit words in the form DCBA. After a short period of getting accustomed to this, it becomes easy. It is doubtful that anyone reads a long binary number as readily as he or she does a comparably long decimal number.

In working with some digital circuits (counters in particular) you have to follow a sequential count in binary form. Thus the binary output of a four-bit counter runs 0000, 0001, 0010, 0011, 0100, 0101, 0110, 0111, 1000, 1001, 1010, 1011, 1100, 1101, 1110, and 1111, which is 0 through 15 in decimal notation. The ability to do this also develops very rapidly.

The Addition of Two Binary Numbers

The set of rules for adding together two one-bit binary numbers is short and simple. The rules are: $0 + 0 = 0$; $0 + 1 = 1$; $1 + 0 = 1$; and $1 + 1 = 10$. The first rule says that zero added to zero gives zero. The next two rules say that the sum of zero and one is one, regardless of the ordering of the addends. The last rule says (in decimal notation) that the sum of 1 and 1 is 2. However, since the binary representation of the decimal digit 2 requires two bits, the binary sum $(1 + 1)$ requires a two-bit binary number. (Similarly the sums of some one-digit decimal numbers require two digits for their expression. Thus $9 + 8 = 17$.)

We turn next to the problem of adding together longer binary numbers, using a specific case to illustrate the procedure. Let us say that the sum of 101 and 111 is to be calculated. The process is shown step by step in the Table A.1. In part (a) the rightmost bits have been added and their sum written below the line as a *sum* bit 0 and a *carry* bit 1, identified as S and C.

Table A.1 Addition of two three-bit binary numbers.

1	0	1		1	0	1		1	0	1		1	0	1
1	1	1		1	1	1		1	1	1		1	1	1
	1	0		1	0	0	1	1	0	0	1	1	0	0
	(C)	(S)		(C)	(S)		(C)	(S)		(Final		sum)
	(a)			(b)			(c)			(d)		

In Table A.1 (b) the three bits in the middle column of part (a) have been added together. The original 1 and the carry 1 in part (a) give a sum bit 1 and a carry bit 1 in part (b). At this stage we know that the rightmost bit in the final sum will be 0 and the next rightmost bit will be 0. To complete the addition, the three 1's in the leftmost column in part (b) must be added together. The sum of the three 1's is 11, the corresponding sum in decimal notation being $1 + 1 + 1 = 3$.

This is indicated in part (c) by showing a sum bit 1 in the third from the right column and a carry bit 1 to the left of that. The carry bit 1 in part (c) becomes the leftmost bit in the final sum, which is 1100 as shown in part (d). In this example the addends are 5 and 7 and the sum is 12 in decimal notation.

This is quite analogous to the way in which you add two decimal numbers. For example, when you add 97 and 15, you say that 7 plus 5 is 2, carry one and then you add the 9, the 1, and the carry 1 to get 11. Hence the sum is 112.

The Representation of Negative Numbers

There are various ways in which the binary form of a negative number can be written. In one method the binary number is preceded by a *sign bit* to indicate whether the number is positive or negative. The sign bit 0 is used for positive numbers and the sign bit 1 is used for negative numbers. For examples, 0110 and 1110 are read as + 110 and −110 respectively, or as +6 and −6 in decimal notation. The reason for using sign bits rather than the symbols + and − is that in digital circuitry there can be only binary digits.

The next method we discuss uses *signed one's complements*. By the complement $\bar{A}$ of a bit A we mean $\bar{A} = 1$ if $A = 0$ and $\bar{A} = 0$ if $A = 1$. The one's complement of a multibit number is formed by writing out a string of bits, each of which is the complement of the corresponding original bit. Thus the one's complement of 110110 is 001001. If the string is prefixed by a sign bit, the result is the signed one's complement of the original number. Thus 0001001 is the representation of +110110 in this type of notation.

In digital electronics there are important technological considerations such as speed and simplicity which make still another scheme preferable to either of the foregoing schemes. The *two's complement* representation is the system in most common use. The two's complement of a number is the one's complement with 1 added to it. Thus the decimal number +10 is in signed binary form 01010, the one's complement of that is 10101, and the two's complement is (10101 + 1), or 10110. Similarly, the decimal number −10 becomes 00110.

Subtraction of Binary Numbers

To subtract +5 from +23 is to add −5 to +23. Similarly the subtraction of a binary number from another binary number can be regarded as the addition of a negative binary number to the other binary number. In this section we will show how subtraction is done using the two's complement representation of binary numbers.

The applicable rule is that a positive number A is subtracted from another positive number B by adding to the binary representation of B the two's complement of the number A. For example, to subtract +5 from +23 according to this rule we first write +23 (decimal) as 010111 (binary) and +5 (decimal) as 000101 (binary). Then we take the two's complement of 000101, obtaining as the result 111011. Finally we add together 010111 and 111011. The direct result is 1010010. However, in this method *the carry which results from the sign bit addition is disregarded.* The tentative result 1010010 is accordingly shortened to 010010. You read that as "positive 10010" in binary form or simply as +18 in decimal form.

Related subtraction problems can arise. An example is the subtraction of +5 from −23, and another is the subtraction of −5 from −23. Such problems can be handled by the two's complement method, but we will not pursue these matters here.

Multiplication and Division

The rules for multiplying together two one-bit numbers are $0 \times 0 = 0$; $0 \times 1 = 0$; $1 \times 0 = 0$; and $1 \times 1 = 1$. The simplicity of these rules leads to great simplicity in multiplying together longer binary numbers. As an example, the multiplication of 110111 and 100101 is shown below.

```
      110111
      100100
      ──────
      000000
      000000
      110111
     000000
    000000
   110111
   ─────────────
   11110111100
```

The rules for division are $0 \div 1 = 0$; and $1 \div 1 = 1$. Division by 0 is undefined. In the process of division of decimal numbers familiar to everyone, you must select a trial quotient digit at each step. This is also true in binary division, but the selection is simplified since the choice of 1 or 0 is dictated according to whether the divisor is equal to or smaller than or larger than the partial dividend. We show how 10011 is divided by 101.

```
           011 Quotient
101   10011
      000
      1001
       101
      1001
       101
       100 Remainder
```

How to Convert a Decimal Number into Its Binary Form

We began this appendix with the conversion of binary numbers into decimal form. We conclude with the problem of conversion from decimal to binary. *This* conversion is carried out by an algorithm, which we state without proof. The given decimal N is divided by 2, with quotient Q_1 and remainder R_1. Q_1 is next divided by 2 with quotient Q_2 and remainder R_2. Q_2 is then divided by 2 with quotient Q_3 and remainder R_3. The process continues until it depletes itself. Then the remainders R_1, R_2, R_3, and so on are written right to left. The binary number which results is the representation of the original decimal number N.

An an example the conversion of the decimal number 86 into its binary form will be shown.

	43		21		10		5		2		1		0
2	86	2	43	2	21	2	10	2	5	2	2	2	1
	86		42		20		10		4		2		0
	0		1		1		0		1		0		1

The result is that 86 (decimal) = 1010110 (binary).

Appendix B:
The BCD and Other Codes

The Multiplicity of Codes

Digital apparatus operates by switching between two voltage levels. These are associated with the binary digits 0 and 1. If a string of 0's and 1's is read as a binary number (Appendix A), a particular interpretation of the string has been made. Thus by agreement 11010 represents the decimal number 26 in binary form. However, it is also necessary in digital electronics to represent the letters of the alphabet and other symbols as well as numbers so that *alphanumeric* information can be stored and manipulated. Suppose that we were to agree that 00000 represents A, 00001 the letter B, 00010 the letter C, and so on. Then the string 11001 represents the letter Z. How could some electronic equipment tell whether 11001 means "25" or "Z"? We might append at the right of any string of bits a rightmost bit which would mean "number" if it is 0 and "letter" if it is 1. Then 110010 and 110011 would mean 25 and Z, respectively.

What we have done here has been to invent a *code,* or set of rules, by which a string of bits is to be interpreted. It might turn out that this code is not as good as some other, possibly because it is not as conservative of storage space or because it does not permit the hardware to operate as fast. It should not be surprising that numerous codes exist, given the variety of applications of digital techniques.

Such codes are all *binary codes.* In this appendix we describe some binary codes selected because they are of fundamental importance in digital electronics (the BCD code), because they are extremely common even though somewhat arbitrary (the ASCII code), or because you may encounter refer-

178

ences to them (the excess-3 and Gray codes) even though you may not have direct need for them.

The BCD Code

BCD means *binary-coded decimal,* although "binary coded decimal digits" would be a better name. In this code a set of four bits is used to represent each of the decimal digits 0 through 9. For a longer set of decimal digits, such as 145, four bits are used for each of the decimal digits. The rule is that the four-bit numbers 0000, 0001, 0010, 0011, . . . , 1001 represent the decimal digits 0, 1, 2, 3, . . . , 9 respectively. Thus the BCD representation of the decimal number 145 is 000101000101. As another example 1982 becomes 0001100110000010. The interpretation of a BCD number is made by breaking the BCD number into sets of four bits, and reading each set as a decimal digit.

The BCD code is used very much; for example in the 7490, 7493, and other counters discussed in Chapter 8.

A BCD number is not the same as a binary number in general. For any decimal digit 0 through 9 the two *are* the same. For example, the decimal digit 7 is 0111 in BCD and in binary. However, the decimal number 10 in BCD form is 00010000 while the binary form is 1010. If the string 00010000 was interpreted as a binary number, it would represent the decimal number 16. Arithmetic does not work the same way in BCD as in binary either. Arithmetic operations in a BCD digital system must be handled by special techniques.

Notice that a four-bit symbol can represent 16 quantities, rather than just 10. The BCD code is a truncated version of a modulo-16 code. Table B.1 shows

Table B.1 Seven-segment display characters shown when driven by a 7447 or 7448 decoder.

BCD Symbol	Modulo-16 Symbol	BCD Symbol	Modulo-16 Symbol
0000	0	1000	8
0001	1	1001	9
0010	2	1010	⊏
0011	3	1011	⊐⌐
0100	4	1100	⌴
0101	5	1101	⊏̲
0110	6	1110	⊢
0111	7	1111	(BLANK)

The horizontal line following the symbols 1001 and the corresponding 9 indicate where true, or *decade,* or *modulo-10* BCD leaves off.

how the full capabilities of four-bit symbols are made use of in the case of some popular digital devices such as the 7447 and 7448 decoders.[1] (See Chapter 8.)

The Gray and Excess-3 Codes

The Gray and Excess-3 (or XS3) codes are shown in Table B.2. The Gray code is of interest because when sequential counting is done in this code it is never necessary to change more than one bit at each step. The Gray code is an *unweighted* code, because the position of a bit is not associated with a power of 2 or some other numerical weight, as is true in the case of the usual BCD code, which is weighted 8, 4, 2, 1.

Table B.2 Gray Code and Excess-3 code

Decimal Number	Gray Code Symbol	Excess-3 Code Symbol
0	0000	0011
1	0001	0100
2	0011	0101
3	0010	0110
4	0110	0111
5	0111	1000
6	0101	1001
7	0100	1010
8	1100	1011
9	1101	1100
10	1111	
11	1110	
12	1010	
13	1011	
14	1001	
15	1000	

The Excess-3 code representation of a decimal digit from 0 through 9 is formed by adding 3 to the digit and expressing the resulting number as a four-bit binary number. Thus 8 is represented by increasing 8 by 3 to get 11 for the sum, then writing decimal 11 as 1011 in binary. This code uses only 10 of the possible 16 combinations of four bits. If one of the unused binary symbols (such as 1111) was to occur in a digital system, that would constitute an error.

[1] The symbols are shown as they would appear on a seven-segment display unit. In other applications the decimal numbers 10 through 15 are represented by the letters A, B, C, D, E, and F, respectively. This is especially true in some microcomputers.

The ASCII Code and Related Codes

ASCII stands for *American Standard Code for Information Interchange*. This code is in widespread use in computer systems and in printers such as the teletype machines produced by the Teletype Corporation. ASCII is a code which uses the bits 0 and 1 in strings to represent decimal digits, both upper and lower case letters, some other symbols such as !, ", #, (,), and control function signals as well. ASCII is basically a 7-bit code, but sometimes an eighth *parity* bit is added. In the *odd parity* scheme the parity bit is chosen so that the sum of all the bits is odd and in the *even parity* scheme the parity bit is such that the sum of all the bits is even. The presence of the parity bit aids in detecting some (but not all) errors in the transmission of data. This is an instance of *error-detecting codes*.

We do not present the entire ASCII code here because the full code has no application within the bounds of this book and because it includes numerous special control symbols which cannot be explained adequately without devoting considerable space to the topic. Table B.3 gives the seven-bit ASCII representations for some characters.

Table B.3 Part of the ASCII code.

ASCII Code	Character	ASCII Code	Character	ASCII Code	Character
0100001	!	1000001	A	1100001	a
0100010	"	1000010	B	1100010	b
0100011	#	1000011	C	1100011	c
0100100	$	1000100	D	1100100	d
0110000	0				
0110001	1	. . . .		. . . .	
0110010	2				
0110011	3	1010000	P	1110000	p
0110100	4	1010001	Q	1110001	q
0110101	5	1010010	R	1110010	r
0110110	6	1010011	S	1110011	s
0110111	7				
0111000	8	. . . .		. . . .	
0111001	9	1011000	X	1111000	x
0111010	10	1011001	Y	1111001	y
0111011	11	1011010	Z	1111010	z
0111100	12				
0111101	13				
0111110	14				
0111111	15				

EBCDIC (Extended Binary Coded Decimal Interchange Code) is similar to ASCII. It is used by specific manufacturers of digital equipment. It is an 8-bit code in which two four-bit BCD words are used to represent each character! is 01011010, 1 is 10001111, and A is 10000011 in EBCDIC.

The *Hollerith Code* is much used with punched card devices. The punched card has 80 columns, each with 12 positions in which a punched hole may be made. For example, if there are holes in rows, 2, 8, and 11 the symbol ! is represented; if there is a hole in row 1, the number 1 is represented; and if there are holes in rows 12 and 1, the letter A is represented.

The *Baudot* code is used in some data communication equipment. It is a double five-bit code. The first five-bit symbol indicates whether the following five-bit symbol is to be interpreted as a letter or a figure. ("Letter" and "figure" are not taken in their literal senses.) The second five-bit symbol selects the letter or figure. Thus 11011 01101 means !, 11011 being the figure indicator, while 11111 01101 means F, 11111 being the letter indicator.

Appendix C:
What Is Needed
for Basic IC Experimenting

We hope this book will be of value to readers who want to acquire only a conceptual understanding of digital IC electronics, but we must say strongly that building circuits is of tremendous help in mastering the conceptual material. It is also an enjoyable activity. Each time you have before yourself a successfully constructed and properly operated circuit, there is a sense of real understanding and a perception that the ideas in digital IC electronics are not mere abstractions but are about very real, tangible, useful, and interesting objects.

Users of this book who are in a formal school will be fortunate in having laboratory set-ups provided for them. In this appendix we seek to help the more isolated readers who have to equip themselves for hands-on experience with ICs.

Power Supply

For work with TTL ICs the power supply should be a regulated supply that can furnish +5 volts DC to within +0.2 volts and can also furnish substantially more than the largest current to a load that you are actually going to demand of it. In our student laboratories we use regulated power supplies which are rated as ½-ampere supplies. These have proven themselves to be fully adequate, although they are not up to the job of powering (say) a basic microcomputer board along with an auxiliary memory board.

183   CMOS chips are more tolerant about their power supplies than are

TTLs. Giving them +5 volts within close limits is not necessary and the current demands made by CMOS are small. However, the power supply described for TTL can be used with CMOS, and there is ordinarily no need to use two different voltage sources.

A suitable power supply can be bought outright. They may be available locally and they are advertised widely by mailorder houses. You can buy a copy of an electronics magazine such as *Radio-Electronics, Popular Electronics,* or *Elementary Electronics* from a newsstand to get numerous names and addresses of dealers. A good +5-volt DC regulated supply should cost less than $30. (This is the most expensive single item you need.)

Readers with some experience in building circuits from scratch could build their own power supplies without difficulty. The circuit shown in Figure C.1 is quite standard and will serve well. An inexperienced reader (or someone who wants to avoid having to collect component parts piecemeal) can easily make a power supply from a purchased kit.

For small-scale experimentation it is possible to use dry cells or batteries. Refer to the first section of Chapter 5, where two ways of doing this are shown. This is the ultimate in economy and simplicity and it will work within limits.

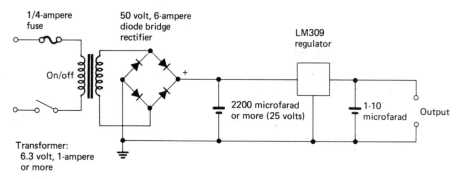

Figure C.1   A 5-volt regulated power supply for TTL and CMOS experimentation.

Solderless Breadboards

An IC *breadboard* is an absolute necessity. This is a plastic panel in which there are numerous holes interconnected in certain ways under the plastic. With a breadboard, circuits can be quickly hooked up and modified easily. Figure C.2 is a drawing of a typical breadboard of the kind which is sold by Radio Shack at their numerous outlets, and by Continental Specialties Corporation, 44 Kendall Street, New Haven, CT 06512, and other dealers. Each circle in the drawing represents a hole into which a wire lead from an electrical component can be

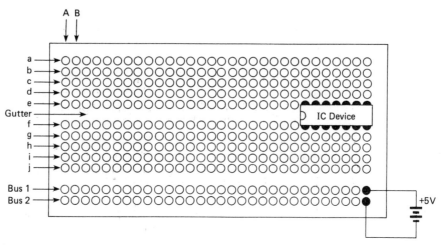

Figure C.2   A solderless breadboard for building IC circuits.

inserted. Holes a, b, c, d, and e are connected in common by a metal strip which is buried in the plastic. These five interconnected holes are not connected to any other holes in the device. Holes f, g, h, i, and j are similarly connected to each other, but to no other holes. Thus the column A consists of two sets of holes, with five mutually connected holes in each of the sets. These sets of five straddle a somewhat wide gutter. Similarly, column B consists of five holes in a vertical row above the gutter, connected together, and five holes below the gutter, connected together, and similarly for the rest of the columns of the device. The horizontal row labelled Bus 1 consists of 30 holes, all of which was connected together under the plastic, and similarly for Bus 2. The usual use for the buses is indicated at the lower right. All the holes in Bus 1 are connected to the positive terminal of the power supply, and all the holes in Bus 2 are connected to the negative end of the power supply, which is ordinarily taken to be the circuit ground, or common. To illustrate how the breadboard accomodates an IC device, a 14-pin chip is shown inserted at the right, straddling the gutter. The half-holes blackened-in indicate where the pins of the device make contact with their respective vertical sets of holes. Notice that for each pin of the device there remain four holes by means of which other wire leads can make contact with the pin. For example, if the +5-volt and ground pins of the device are pins 14 and 7, respectively, short wire jumpers can be run from any hole in the Bus 1 set over to any of the four holes that connect with pin 14, and from any hole in the Bus 2 set over to any of the four holes that connect with pin 7. This would provide power to the device.

Breadboards are available in both smaller and larger sizes than that shown. The smallest will hold only one or two ICs and the largest may hold more than a dozen. Many have flanges along the edges which make it possible to cascade small units into a larger unit. Prices range from a few dollars up to more than $20, but you will never regret your investment in basic breadboards. There can hardly be any IC experimentation without them.

There is one more detail of which you must be aware. The holes in breadboards of the kind we are describing accept number 22 bare solid (not stranded!) wire ends. Number 24 wire can be used in an emergency but the fit of the wire to the holes is not really tight enough for satisfactory connections. The leads from resistors or other circuit components may be too thick, and you must not force such heavy leads into the holes. Instead you should either use a component with thinner leads or solder pieces of number 22 wire onto the existing leads.

Logic State Indicators

When you study the behavior of an IC device you need some way to determine whether the state of any pin is high or low. With very low-speed experiments, such as those in which you manually operate a switch to change the logic states of pins, the state indicators can be very simple. At high rates a cathode-ray oscilloscope is needed.

A suitable indicator can be a simple DC voltmeter reading to 5 volts. A high input impedance VOM (volt-ohmmeter) or VTVM (vacuum tube volt-meter) would be better, to avoid having the meter load the circuit.

A good and economical logic state indicator is a discrete LED. (See Chapter 9.) An LED with its cathode connected to the ground bus bar of a breadboard and its anode touched to an IC pin will light up if the pin is high and will remain off if the pin is low. Notice that an LED may require a current-limiting resistor in series with it. Usually this resistor is 330 ohms or so.

Devices called *logic probes* are available on the market. In one form, such a probe indicates by some kind of readout whether the pin to which the probe tip is touched is at logic low, logic high, or pulsing. It is a simple matter to make a good and inexpensive probe. A suitable circuit is shown in Figure C.3. This probe will work with either TTL or CMOS chips.

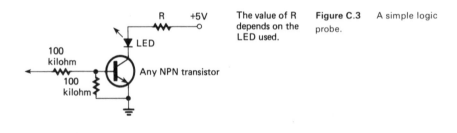

The value of R depends on the LED used.

Figure C.3   A simple logic probe.

A Seven-Segment Display Unit

Digital electronic systems commonly use seven-segment display devices to present their outputs in the form of decimal digits. You will need at least one seven-segment display unit. If you are performing experiments as we suggest them in this book, your first real need for a seven-segment display unit will appear in connection with Chapter 8. When you are ready to equip yourself with a display, Chapter 9 should be referred to for technical help.

Debounced Switches

In many experimental circuits it is necessary to change the states of some IC pins manually. Sometimes this can be done simply by pulling the appropriate wire end from, or inserting it into, the +5-volt bus or the ground bus, as may be needed. You can, of course, use actual switches but this tends to make the wiring of simple circuits unnecessarily elaborate. Either method will work with simple gates and even with more elaborate combinatorial circuits.

However, due to switch bounce neither method is likely to work satisfactorily when counters, flip-flops, or some other devices are involved. Switch bounce is discussed in Chapter 4, and there a simple method to make a debounced switch unit is given. You should have at least a pair of debounced switches. One pair requires only one gate package, two singlepole double-throw switches, and some wire.

The Integrated Circuits Themselves

TTL, CMOS, and other integrated circuit devices are available from a large number of mailorder houses and from local across-the-counter dealers. A copy of an electronics magazine will give you numerous names and addresses of mailorder dealers.

Many ICs are priced at less than the cost of a postage stamp. Many more cost less than one dollar; only a few cost more. For general experimentation the cost of the ICs needed is truly small and this in spite of the fact that the ICs are precisely the objects you are studying!

Hand Tools

The hand tools needed are only three in number. You need a pair of side-cutting pliers (often called diagonal cutters). They are used to cut wire to length and strip off insulation. Much more expensive wire-strippers can be purchased but with a bit of practice you can learn to strip insulation with side-cutters just as well and even more quickly.

You also should have a pair of small needle-nose pliers. These are of great help in pushing wire ends into the sockets of a breadboard, for straightening IC pins, and for general manipulation of small objects.

Finally a screwdriver with a blade ⅛ inch or so in width is useful. A screwdriver is, of course, useful in many ways but it is noteworthy that it is a great aid in prying ICs loose from breadboard sockets. Special instruments (IC pullers) can be bought but they are not really necessary for most experimental work.

Electrical Components

Resistors and capacitors are not needed for the most part in basic experimenting with ICs. The experimenter is most likely to need them only in making a 555 clock or some monostable circuit or in association with LED devices. In any event, they are easily obtained.

We have already mentioned connecting wire. It should be number 22 solid insulated wire, neither stranded or bare if standard breadboards are to be used.

An item which can be very handy is a set of simple SPST (single pole, single throw) switches in *dual in-line* form. These DIP switches come four or eight to a package. The package has the general form of an IC and the pins fit directly into an IC breadboard. Their highly compact form helps very much in avoiding having large off-the-board components with long runs of wires.

The last item of hardware which we recommend is an item which can solve a great many interconnection problems in building temporary circuits. It is the alligator clip. Suppose you need to use a resistor which has leads too thick to enter a socket in the breadboard. A short piece of number 22 wire with an alligator clip attached to one end takes care of the difficulty. The bare end of the wire goes into the appropriate socket and the clip bites onto the resistor lead. A small alligator clip costs only ten cents or so and usually six are sufficient.

Lab Manuals

At the ends of many of the chapters in this book you will find suggestions for experiments. Most of these are described very briefly, deliberately. The reason is that it is desirable that you go to the pin diagrams for the ICs you will use and draw up your own circuit layouts. Nevertheless, we recognize that an experimenter may want more complete circuit diagrams and help from accompanying commentary. In short, you may want to get a lab manual.

The most popular published books of experiments are the volumes in the *Bugbook* series. Those numbered beyond I and II are on specialized topics, but *The Bugbook I* and *The Bugbook II* (both subtitled "Logic and Memory Experiments using TTL Integrated Circuits") are just what the learner needs. These are written by Peter R. Rony, David G. Larsen, and Robert A. Braden and are published by E & L Instruments, Incorporated, 61 First Street, Derby, CT 06418. These volumes contain much text material and give elaborate experimental instructions.

A brief summary of the topics in *Bugbook I* is: breadboarding, gating, truth tables, counters, decoders, multiplexers, and sequencers. For *Bugbook II* it is: LEDs, three-state and open collector outputs; flip-flops and monostables, RAMs and ROMs, shift registers, arithmetic elements, and Schmitt triggers.

Another good lab manual is *Digital Experiments,* second edition, by Richard E. Gasperini, published by Movonics Company, Box 1223, Los Altos, California 94022, in cooperation with the Hayden Book Company, Inc., 50 Essex Street, Rochelle Park, NJ 07662, 1978. This is a very useful smaller manual (about 200 pages), containing 25 experiments ranging from "Familiarization" through LED displays, decoders, multiplexers, flip-flops, counters, memories, and more.

Index

191